MY WILD KINGDOM

MARLIN
PERKINS

MY WILD
KINGDOM

AN AUTOBIOGRAPHY

E.P. DUTTON, INC. | NEW YORK

Published in the United States by E.P. Dutton, Inc.,
2 Park Avenue, New York, N.Y. 10016

Library of Congress Cataloging in Publication Data

Perkins, Marlin.
    My wild kingdom.

        1. Perkins, Marlin.   2. Zoologists—United States—
Biography.   I. Title.
QL31.P43A35   1983   590'.92'4   [B]   82-5102
                                                       AACR2

ISBN: 0-525-24146-9

Published simultaneously in Canada by
Clarke, Irwin & Company
Limited, Toronto and Vancouver

Designed by Earl Tidwell

10 9 8 7 6 5 4 3 2 1

First Edition

To my wonderful wife, Carol,
and to our family,
Mary Ellen and Fred, Alice and Ren,
Suzanne, Marguerite and Peter

Contents

Photographs follow pages 40 and 180.

MY WILD
KINGDOM

1

A Missouri
Boyhood

At the beginning of spring in 1905, on March 28, I came into the world in my parents' house at 902 South Main Street, Carthage, Missouri. My brother Leland preceded me by nearly seven years and my brother Carlyle by three and a half. Each of us was born in our mother's bed, as was normal in those days, but she did have a woman physician, Dr. Ketchum, in attendance.

Our branch of the Perkins family emigrated to Virginia in the 1600s when one of the three brothers from England settled there, in Albemarle County. Another brother settled in the Boston area, and the third brother somewhere in Georgia or South Carolina. After the Revolutionary War, when my grandfather was one year old, my great-grandfather moved with his family to Shelby County, Kentucky. After eighteen or nineteen years there they moved on to Missouri, where my father, Joseph Dudley Perkins, was born in 1851 on a small farm three miles from Farmington. His father, Isaac Hardin Perkins, besides being a farmer, was also a carpenter. He had married Elizabeth Horn, a woman of Dutch descent. Both my paternal grandparents died before I was born.

My mother's family, the Millers, were German-speaking Alsatians, and they progressed westward through the United States from Pennsylvania to Ohio and Kansas, where my mother, Mynta Mae Miller, was born at a little village called Osage Mission—a town no longer in existence. She was the youngest of seven children. Her father made two or three trips across country from Missouri or Kansas to Carson City, Nevada, in wagon trains carrying loads of merchandise to be sold when he arrived. He told me that late each afternoon the travelers would find a place to stop and station all the wagons in a circle for the night. The axles of the wagon wheels often had to be greased. One night Grandfather watched as two fellows were straining to lift one corner of their wagon. Try as they might, they couldn't get the wheel off the ground. Grandfather Miller was a very strong man. He brushed the struggling men aside, backed up to the tail end of the wagon, lifted both wheels off the ground, and held the wagon end up until both the axles were greased.

By the time I knew him, however, Grandfather was crippled with rheumatism and had to walk on crutches. He had a white beard and smelled of tobacco; when he came to visit us he always brought a red-white-and-blue-striped candy bag filled with marshmallows covered with browned coconut. In his later years Grandfather was in the real-estate business in Carthage.

My earliest recollection of a family dinner with our grandparents was of a spread of chicken and noodles with sumptuous quantities of various vegetables, followed by apple dumplings for dessert. The meal was consumed in two shifts by successive groupings of aunts, uncles, and cousins; the noneating group listened to the Victrola with its morning-glory horn, squeakily playing "My Pretty Red Wing" and other songs, or ran from room to room to watch the hours strike on the twelve or fourteen cuckoo clocks that somehow never went off at the same time. When our grandparents spoke German, we all knew they had a secret and didn't want the kids to understand.

As a young man, my father often went to Farmington when the court was in session. He would climb the stairs to the courtroom and sit entranced by the eloquence of the lawyers and the judge. After trying railroading, farming, and carpentry, my father decided to become a lawyer and made arrangements with Mart Clardy to read law in his law office. After a year or so, feeling he was then qualified, my father took the bar examination in a courtroom of Saint Genevieve County and was examined by the lawyers present. He passed the examination, became a lawyer, and hung out his shingle. Through a series of circumstances he became prosecuting attorney for a term or two, but finally decided that the Farmington community was too small for his future; and he left there to find a new location. He made a trip to Nebraska and liked it pretty well there—well enough to apply for and receive a license to practice law in that state. He didn't stay in Nebraska, however, but came back south to Kansas City, Missouri, and from there to Carthage. It must have been love at first sight, because he settled in Carthage and made it his home for the rest of his life. He was eventually elected judge of the circuit court, division 1, of Jasper County—a position he held for twenty years. The fact that he was reelected to this judgeship for such a long span of years as a Democrat, in a Republican community, speaks well for the esteem in which he was held in the county.

Father came home from court one day to find Mother and me in the bathroom. I was being undressed for a bath, as I had fallen into an excavation for a new house that had been filled with rainwater. I was clinging to a floating piece of wood when I was rescued by a neighbor who had been alerted by our dog running up and down the bank and barking wildly. I was only about two, and don't remember it at all, but Father described it as a very muddy little boy getting a thorough scrubbing. Ever since I have been fond of dogs.

I remember my life as a small boy as a time of almost constant activity. Suspended from a limb of a large tree in our yard was a bag swing—a gunnysack filled with straw, tied at the top, and dangling from a single rope. From the back porch I could jump onto this bag, wrap my legs around it, and swing out over the alley. A favorite feat was to maneuver my position so my legs were toward the porch and my brother Carlyle could jump onto my lap on my first swing back. One day the rope broke, and my brother came down on top of me.

Another activity in our yard depended on a very heavy wire, one end of which was wrapped around a tree at a height of about fifteen feet from the ground; the other end was attached to the trunk of another tree at about four feet from the ground. Riding on this wire was a trolley arrangement consisting of a pulley from which hung a wire with a loop at its lower end; a short section of a broom handle was thrust through the loop. A platform was built up under the first tree, and with a hooked pole you could draw the pulley up to you, grab hold of the broomstick, jump into space, and ride down the wire. Usually you hit the ground running; but a popular variant was to pull yourself up so that you could wait until you braked to an abrupt stop with your feet against the other tree.

Our yard was a fairly small one, so Father bought a piece of property behind ours which had a deep backyard. He rented the house, with the proviso that we kids could play in the backyard. Another piece of our play equipment was an a six-inch-square timber firmly embedded in the ground. Driven into the top, about four feet above the ground, was an iron rod an inch in diameter, threaded so that a nut could be screwed onto it to hold the sixteen-foot-long plank that became our teeter-totter and merry-go-round. Handles a couple of feet from the end of the plank kept us from being thrown clear off. Two of us would sit on the ends of the plank; a third person would stand near the center and start the plank rotating; when it was going lickety-split, he'd run out between the rotations and the riders would have a nice long trip. When a big boy developed a high-speed rotation, the centrifugal force was so great that by relaxing the grip of his legs and hanging on only by his hands, a rider could extend his body into space parallel with the ground.

Another home-built piece of equipment for exercising (or getting killed) was made from planks about ten inches wide by two inches deep and about twelve feet long, attached together, and supported by wooden bases like a trestle. This slide started from a platform in our backyard at an elevation of about fifteen feet, crossed the alley into our other backyard, and finally descended to ground level. Down the center of that long series of planks was nailed a continuous strip of wooden lathing. A riding car was constructed, about three feet long, also made of ten-by-two-inch planking. Two wooden laths were nailed to the car's bottom so they fit loosely enough over the single strip to ride the track. Its bottom properly lubricated with axle grease, the car raced down the incline, providing a very exciting adventure. Surprisingly, the car seldom actually jumped the track.

My brothers had bicycles, and before I could even ride one, they were carrying me around on their handlebars. When I was about three years old Santa Claus brought me a tricycle. Having my own means of transportation thrilled me, as I now could explore the neighborhood. I could see, however, that a tricycle had speed and maneu-

verability limitations compared with a bicycle. My legs were too short
to fit my older brothers' bikes, but I tried to ride by putting one of my
legs through the triangle formed by the frame with one pedal at the
top of its rotation. I'd hang on to the handlebars, push mightily, and
ride the bike in a cant from the vertical. The system worked, but it
was not a satisfactory way to ride a bicycle. Realizing this, my brothers converted my tricycle into a bicycle by taking off the two rear
wheels, removing the axle, and squeezing the sections of frame close
together. One of the rear wheels was then repositioned on a short axle
between the frame sections. After a little help from a steadying hand
on the back of my seat, I soon learned how to balance while moving
forward. It wasn't a real bicycle, but it was faster than the tricycle
and safer than a bike in turning a corner.

When I was a small boy we had a wonderful long-haired black-
and-brown shepherd dog named Rex. He was an outside dog and was
never confined, but was a wonderful pet and a fine watchdog. Our
father bought us a billy goat, white in color, with part of his horns
removed. He was kept tied to a tree with a long rope, and one of our
jobs was to make sure he didn't get tangled up. He was trained to pull
a little farm wagon with three seats in it. With Billy hitched to the
wagon in shining harness, we were the envy of all our friends as we
sportingly drove around our neighborhood. Mother and Father made
some rules about where we could and could not go with our billy goat
and wagon. Alleys and side streets were safe enough, but we were not
allowed to drive along Main Street because of the streetcars. We
could cross Main Street, but only when no streetcar, horse and buggy,
or automobile was in sight.

There was an interurban electric streetcar that connected Carthage with Joplin, Missouri—going through Carterville and Webb
City on the way. It circled the square in Carthage and ran right down
the middle of Main Street, passing our house. All you had to do in
those days was tell the conductor where you wanted to get off, and
he'd stop there. When holding court in Joplin, my father would ride
the streetcar back and forth. Our dog Rex had a game with these
streetcars. He would lurk behind a tree, and as one approached, he'd
dash out, barking ferociously, and chase it down the street. Being a
fast runner, he could frequently keep up, and once in a while he
would grab hold of the bottom wooden step with his teeth and hang

on. This game became a great annoyance to our family, particularly to my father, who rode the car so often that he knew all the motormen and conductors and was aware of their opinion of that obnoxious dog. Knowing that Rex was gun shy, my father decided to put a stop to the dog's antics. He gave a blank pistol to one of the conductors and asked him to fire it near the dog when Rex grabbed hold of the step. Greatly shocked, Rex stopped chasing streetcars for a while, but the problem was never really solved until we moved from 902 South Main to our country home three miles out of town.

Father and Mother had bought ten acres on a little hill, and there they built a very nice house. The first floor was constructed of white limestone rocks, part of the residue from the lead and zinc mines nearby. Some had calcite crystals in them and were locally known as tiff. My mother personally selected outstanding examples of this tiff rock for the living room fireplace and for the vestibule entrance of the front door. The second floor of the house was of brown shingles, and there was a sleeping porch clear across the south side of the house. Below this, just off our dining room, was a solarium with a concrete floor and a drain in the lowest spot; this was where mother kept her plants.

The streetcar line ran in front of the house, so Father could ride into either Carthage or Joplin, depending upon where circuit court was being held. Our ten acres included a nice apple orchard and several fields, a barn, a big toolshed, and a little house that Father moved to the back of the property as a residence for his hired hand. The hired hand farmed the place, and his wife helped Mother in the kitchen and with the washing. There was a big garden, and we always had plenty of vegetables. There was also a patch of peanuts. One day I was with our hired hand when he was digging the peanuts. He wiped the dirt off one, broke it open, and suggested I eat it. What a surprise to learn that it didn't taste like a peanut! Instead, it was soft and tender and a little bit sweet.

Power from a windmill kept us supplied with water, pumped into a huge tank in the basement for household use, and into the barn's tank to water a cow and our horse, Dan.

Dan was the kind of farm horse that could plow, pull a wagon or buggy, and also be ridden by small boys. My two brothers and I used to ride him bareback. Leland rode in front because he was the largest; then came Carl, who held on to Leland; and then myself, holding on to Carl. Dan's back was so broad we couldn't get much of a grip with our legs, but Leland clutched Dan's mane to keep us from sliding backward. One day we were riding Dan in a large pasture near our farm. The grass was high and lush, and daisies and Indian paintbrush were scattered among the tall grass. The breeze blew waves and ripples through the sweet-smelling grass. Leland urged Dan to move a little faster, and before I knew it the horse was in a full run. The wind whistled past my ears, and I held on to Carl with all my strength. We didn't see it ahead, but all of a sudden Dan leaped in a long arc across a small brook, and then we all saw the fence not far beyond. Dan braked fast, and quick as a flash all three boys went over his head and landed on the ground. Dan turned to try to keep from stepping on us. Being the last one off, I was the closest to him, and I saw his front leg coming down and felt a pressure on my right hand as his hoof pressed into the soft ground. When my frightened brothers asked if I was hurt, I showed my dirty hand and moved my fingers to prove it was all right. I could hardly believe that the full weight of our big farm horse's front hoof had not injured my hand. Neither could my mother and father.

This farm was a great place for kids and particularly for me because it was full of all kinds of animals. I'd follow the hired hand when he was plowing and watch the creatures that were uncovered by the plow. Grubs, earthworms, beetles, a nest of baby mice, thousand leggers, and once a toad. Many of them went into my pockets, including the baby mice, but mother frowned on this and I was not allowed to keep them. I found a garter snake at the back of the property one day, but it moved so fast and I was alone—and only six years old—so I walked away from it.

We read *Boy's Life* and *Popular Mechanics*, and my brothers were great inventors and builders of things. One day they decided to build a glider and got together all the necessary materials. Following the design in one of the magazines, they built a contraption which they covered with cloth signs that had been used as advertisements on barns in the neighborhood. Mother sewed the signs together so they would fit. When all was ready, Leland hopped off the roof of the barn, achieved only a small lift, and crash landed about twenty feet away. On another occasion, my brothers built a huge kite. When it was ready to fly, they ran against the wind to lift it. It flew so erratically that they quickly let it come to the ground and said to me, "Marlin, it needs more weight on the tail. Here, take this stick and put it through that tail rope and hang on." With the next gust of wind, they started running with the kite. I ran with it too, until it picked me up and started toward the sky. I must have sailed up about thirty feet before my brothers panicked and ran toward the kite with the rope to make it come down again.

When I was nearly seven years old, I came down with pneumonia and was very sick with fluid in my lungs. My mother nursed me day and night. Dr. Gentry came to see me every day. He and Father both tried to get Mother to bring in a nurse so that Mother could get some rest. She refused, saying she would rather take care of me herself. In about two weeks she too came down with pneumonia, and in ten days she was dead. The funeral was held in our living room. Dr. Gentry took me out of bed, wrapped me in a blanket, carried me downstairs, and held me looking down into the open casket. He whispered in my ear, "Kiss her on the forehead," and lowered me down. As my lips touched her cold forehead, I heard a moaning gasp from one of those dear old ladies with lace around her neck. I can still hear that gasp.

To lose my mother at age seven was a trauma of the first magnitude. The void left by her departure could never be filled. I remember running around and around our house crying, "I want my mother, I want my mother," as though by hurrying I might in some way catch up with her. At the same time, I knew she was dead and buried in the cemetery. There was a marker for the place and a large stone in white Carthage marble engraved PERKINS. There was also the excruciating conviction that she had caught pneumonia from me and the question whether it was fair that I had lived and she had died. Perhaps, I thought, it would have been better the other way round. But as I grew older I realized that I didn't have a choice in the matter, as individuals never do, and gradually the hurt eased.

Father, twenty years my mother's senior, took a long time getting used to the loss. He employed Mrs. Doshia Kingsbury as housekeeper, and we three boys continued our education. I rode the interurban line into Carthage and then transferred to the Toonerville Trolley type of streetcar that ran from the northwest corner of the square down to the Frisco Railroad Station. This passed within a block of my school, and school was only a block from my grandparents' home, where I'd often stop for a crust of fresh homemade bread liberally spread with plum butter or peach jam. Sometimes, though, when I had reached the Carthage square and had twenty minutes or so to wait before the next car left, I'd pop into McCormick's grocery store and charge a small paper bag of Hydrox cookies to my father's account. One day when the bill came in, my father spoke to me about it, and when I explained to him my terrible hunger, he patted me on the head and said, "That's all right, Marlin. If you get hungry on the way home, you just stop and get yourself something to eat and charge it to me." That gave me a wonderful feeling of security.

2

Prairie-Dog Country

In 1912, at the end of the school year, my father bought a Model-T Ford. He was shown how to drive it, and so were Leland and Carlyle. Two weeks later Father took us on a trip to Colorado, going first to Pittsburg, Kansas, for a brief visit to relatives, then on to Fredonia, Kansas, where Father had a nephew named Leo. From there west, the roads were not what we would call highways today, but mostly two tracks with a third track in the middle where the horses pulled the wagons or the carriages. We were constantly stopping to ask directions and covered only about sixty or seventy miles a day—occasionally reaching just over a hundred miles a day.

In western Kansas when we were following the ruts through a field of waving grass, and the wind got so hot it was uncomfortable, the water in our radiator would start to boil. We would stop, get out, unscrew the radiator cap, and then while the water was cooling— before we could pour in more water—we'd have to clean the grasshoppers out of the little square holes in the front of the radiator. Their heads were just exactly the right size to stick in the holes, and it took a lot of doing to dislodge them, even with a wire. As a large percentage of those little square holes repeatedly filled with grasshopper heads, boiling over was a common occurrence. We refilled our water bottle for the radiator every time we could, as well as the Thermos bottle with water for ourselves. We stopped frequently to look around at the countryside or to change a tire.

At one of those stops we saw a whole field of prairie dogs. I couldn't get over how absolutely wonderful they were. They were the cutest little animals I'd ever seen, and the way they'd stand up on their hindlegs, throw their heads back, and give out little barks, made you laugh out loud. Of course, I tried to catch some. An animal would let me get within thirty or forty feet of it, and then, with a quick bark, would dash down a hole. I couldn't see very far down the hole. Then I'd turn and try to sneak up on another one. No use! From there on, we saw prairie dogs in many places, and sometimes a field extending a mile or more would be filled with prairie-dog burrows and prairie dogs running all over.

We came to Pueblo, Colorado, and then drove on up to Colorado Springs, where Father's sister-in-law, named Molly, and two unmarried nieces lived. We visited the Garden of the Gods and Pike's Peak and rode some burros—not very smooth riding. At one place we three boys tried our luck at scaling a sheer cliff of rock, reaching the top of the precipitous wall in first, second, and third place according to age and height. I was a little scared, as the climb seemed steeper coming down than going up, but none of us fell and Father finally quit mopping his brow.

We then drove up to Denver, where Father had a couple of cousins. One, Jim Perkins, was an ex-mayor and still a lawyer in town. The other cousin, G. W. Perkins, was a well-known surgeon. We saw the state capitol with its shiny gold dome and the step with a plaque reading: This Step Is One Mile Above Sea Level. We drove up to Red Rocks and over a pass, and coming down the mountain we burned out a band in our planetary transmission and had to get the system repaired.

My brothers drove most of the time, but occasionally Father would take the wheel. We came back by a different route from the one we'd taken out, and just before we reached home in Missouri, Father nearly wrecked us when he temporarily lost control of the car in some dust. The old Model-T Fords had a strange habit of trying to wrest the wheel away from the driver, and it took a person with some strength to overcome that.

At the end of that school year, Father had decided that the two older boys needed to go to a private school, so they were sent off to Todd Seminary at Woodstock, Illinois; I went to Pittsburg, Kansas, to live with my mother's sister, Aunt Laura, and her husband, Uncle Harvey Black. I was enrolled in the third grade of the Pittsburg Normal School (since called Pittsburg State Teachers College, and now a branch of the University of Kansas). The school was about a mile from where Aunt Laura and Uncle Harvey lived, and in good weather I commuted on my bicycle or roller skates.

One day, Professor Dellinger of the zoology department brought to class a live bull snake about six feet long, named Mike. Dellinger told us a little about snakes while holding Mike, and allowed the snake to crawl from one shoulder to the other and down across his hands. Finally, he said we could all have an opportunity to touch the snake. To my surprise, the animal wasn't anything like what I had expected. He was strong; he had rough scales on his back with little keels down them. His tongue came out of a small hole at the front end of his mouth. His eyes could move underneath a clear crystal of skin, but he didn't have any eyelids and he didn't have any ears. His belly scales were as smooth as smooth linoleum, and you could see where his tail started. He could hang on to you with any part of his body, so that he could keep from falling. I was fascinated. A few days later a couple of friends and I went to the zoology laboratory to see Mike again. Professor Dellinger allowed us to take the snake out of his cage and handle him and then showed us the other live specimens in the collection. Dellinger pointed out a copperhead and told us it was poisonous and that if we saw such a snake outdoors we mustn't go near it, but must *leave it alone.*

Soon afterward, while playing near the school, I happened on a blue racer. With the help of a friend I was able to catch it, and, of course, it bit me. But that didn't hurt too much, and I proudly took the snake to Professor Dellinger. He graciously accepted the gift, and thanked me. It wasn't long before snake catching became a favorite occupation. I soon found places where I could collect certain kinds of snakes or lizards. There was one little area where glass snakes (legless lizards) were abundant. The first one of those I tried to catch, I got by the tail, which immediately broke off in my hand. I thought I'd killed the poor creature, but his tail was wiggling so vigorously I wasn't sure. The rest of the lizard was nowhere to be seen. Later I learned to grab a glass snake near its head, where it couldn't break off its tail.

Excavations for cellars were good places to get the yellowish-red clay so common in our neighborhood. The clay was fun just to squash your toes in and was the right kind of material for a game we invented called mudball whacko. This sticky clay had just the right cohesiveness to be evenly molded with one good squeeze onto the pointed end of a slender, springy willow branch. The clay projectile was launched at the end of a slow-starting but evenly accelerated overhand swing of the willow branch. Just before the end of the swing, a snap of the wrist gave the clayball its final fling.

We started out aiming at the side of a barn. When the clayball hit the barn, it made a sound we described as *whacko*. With a little practice, we became accurate enough to hit predetermined targets, such as the board with some paint missing or the right-hand board in the barn door. Other targets were such obvious things as a cornerstone in the foundation of a building, a tree, and eventually a friend. This last, of course, led to retaliation, and before long we were choosing up sides and having battles much the same as snowball fights in winter.

Overexuberant as most boys become in a battle game, we sometimes got into trouble. Once my friend Maurice King and I decided to see if we could hit some strange kid riding his bike a good three-quarters of a block away. We had to duck out of sight when a clayball connected with his head and knocked him to the ground. Another time, my full-grown cousin Marie, walking along so far away that contact was a real long shot, got her floppy party hat knocked off. I was in disgrace for a week over that.

In our neighborhood we played most of the usual games—hide-and-seek, fox-and-hare, marbles, croquet, horseshoes, empty-condensed-milk-can hockey, baseball, softball, basketball—but a game I have never seen anywhere else was kitty cat. A six-inch length of broom handle was sharpened on both ends, with a long sloping bevel, as in a pencil. Another length of broomstick, with both ends rounded, became the bat. The short piece was placed on the ground. A sharp tap with the bat on the beveled end of the short piece made it spring into the air, and the player quickly hit it a one-handed blow with the bat and sent it flying through the air up the alley toward a line marked in the earth about fifty yards away. Two or more players took turns, and the first to drive the short stick beyond this line got a point. Reversing direction, the winner then batted again, toward the starting line. This game called for some skill and agility—and lots of luck. The short piece did not spring into the air in the same way each time. If lying on a flat surface, it usually rose at least as high as your head— sometimes higher—and that gave you time to follow its arc and position yourself for a good swing. But if it was lying on an uneven surface, the spring into the air was less certain as to height and direction, and your chance of a long hit less likely. If four played, it was more exciting to play partners, then change partners for a second game. The number of points constituting a game could be determined before starting to play.

In the fall of 1919, when I was fourteen and had finished the eighth grade, I entered Wentworth Military Academy in Lexington, Missouri. It was not easy to break away from Pittsburg and to know that I would not be seeing my old friends again on a regular basis. I knew I'd miss Aunt Laura and my other relatives, but several of my friends were also going away to school and that made it a little easier.

At Wentworth I was assigned to C Company and to a room in the barracks called Marine Hall. The room was sparsely furnished with a double-decker metal bed for two cadets, a table and two straight chairs, a book rack and goosenecked reading lamp. Our uniforms were nearly identical with those worn by the infantry in the U.S. Army, with olive drab jacket and trousers flared to the knees, tightly rolled puttees and broad-brimmed campaign hat. Each cadet was issued a Springfield rifle and it was his duty to keep it clean. Inspection could occur at any time.

Discipline at Wentworth was strict and was enforced by both faculty and cadet officers. Hazing kept us "rats" on our toes the whole first year. We had to shine the "old boys'" shoes, clean their rooms, run errands for them, and even light their cigarettes—though smoking was strictly forbidden by the academy regulations and a cadet could be expelled for it if caught by a faculty officer.

One lovely, sunny afternoon in the spring I took a walk a little distance off campus. By the edge of a field I caught two blue racer snakes about five feet long. They were the bluest ones I'd ever seen, so

I let them crawl inside my shirt, where they settled down next to the warmth of my body. Taking them back to my room, I put them in a pillowcase and hid them in an open space behind the plasterboard at the back of my closet. Later I found a nice wooden box that fitted the space and transferred them to that so they'd have more room to crawl around. Daily room inspections failed to disclose the snakes. Each afternoon they were exercised on a secluded lawn in back of our barracks. One day a faculty officer, Captain Hall, came by, saw me with the snakes in my hands, and excitedly ordered me to get away from them. I assured him that my pets were blue racers and were not poisonous, and offered to prove it by letting one of them bite me.

"Perkins," the captain shouted, horrified, "I want you to get rid of those snakes immediately! I've been in India and I know something about snakes!"

He was so upset I thought I'd better take them back where I'd caught them and release them.

I enjoyed my classes. General Science introduced me to some biology, chemistry, physics, geology, and scientific laboratory techniques. Colonel Ovid Sellers, son of the commandant, taught ancient history. He had studied with Dr. Henry Breasted of the University of Chicago and had gone with him to the Near East on digs. He knew Sanskrit and enlivened his classroom lectures with accounts of his experiences. One day, speaking about the Hittites, he mentioned that they were dolichocephalic—had long heads. He wrote the word on the blackboard and then, turning to the class, looked around until he saw me, and said, "Perkins, you are dolichocephalic." He then went on to talk about some neighbors of the Hittites who were brachycephalic—short, round-headed people. He wrote that word on the blackboard, again turned to the class, looked around, and fixing his gaze on another cadet, said, "Adams, you are brachycephalic." Then he said, "There is another word you should know, though I hope there are none of that type in this class—osteocephalic." I knew that meant boneheaded.

Captain Hinton taught English. He was a round, jovial fellow with a resonant voice and excellent diction. As a special treat, about once a week he would read an O. Henry story to us. We all looked forward to that, and I for one have been an O. Henry fan ever since.

It was a good school, Wentworth, but after two years my military career ended when a welcome new development in our family made it possible for me to return home to Carthage.

In 1921, when I was sixteen years old, my father married Laura Gash-weiler, who had been a widow for some time. I had the honor of being best man at their wedding. The old house at Ninth and Main was reopened, and I had two wonderful years there attending Carthage High School, only two blocks away. Mother Laura made a glorious, happy home for Father and me, and for me it was exactly the right thing at the right time. She and her first husband had had no children, and with me as her son she made up for lost time. Her love for me was expressed in many ways, and she was gentle in instructing me in the social graces. One day I walked home from school at noon with a girl who lived a couple of blocks farther on. Mother Laura greeted me inside our front door with a kiss, and then, as we sat down to lunch, she said, "It is proper for a gentleman, when taking his leave from a lady, to tip his hat." I not only accepted such tactful suggestions from Mother Laura but eagerly sought her opinions.

Mother Laura was a good cook, and she was always preparing special dishes she knew I liked, such as fricasseed chicken and home-made biscuits with gravy, strawberry or peach shortcake, or deep-dish banana cream pie. When my friend Lyle Chamberlain and I set out one day on an ill-fated boat trip down the Spring River, Mother Laura had one of these luscious banana cream pies all ready for us to take along. I still hate to remember how we managed to capsize our boat in the middle of the river, treating its population of catfish and snapping turtles to almost all our provisions, including that banana cream pie.

About this time Mother Laura produced one of her most constructive suggestions so far as I was concerned. She proposed that since a young neighbor, Raymond Cozad, and I had birthdays only a week apart, she should arrange a combined birthday party at our house. We invited a large group of friends and classmates. Mother Laura provided punch and lemonade, little sandwiches, and a beautiful big birthday cake, candles and all. After we cut the cake, home-made ice cream was served. We played some parlor games, and I have never had a better time. Before long I was included in lots of other parties.

I joined a debating team at school and was assigned to support

Benjamin Disraeli in a debate on the question of whether he or Gladstone was the greater orator. For reference material, I had only a couple of books from the library and the *Encyclopaedia Britannica*, but Mother Laura gave me a quote she remembered from her own schooldays that fitted in beautifully at the climax of my rebuttal speech. It was Disraeli's crushing reply to Gladstone: "You Sir, are a sophisticated rhetorician inebriated by the exuberance of your own verbosity." I expect it was Mother Laura's quotation that won me the debate.

3

The Joy of Hard Traveling

There was no moon that night, but the stars were bright, and I lay for a long time looking up at the stars and wishing I knew more about them. My thoughts drifted across the sky to Carthage and my father and Mother Laura and my many high-school friends there. Most of my classmates were already scattered in various colleges, but I had taken a year off between high school and college to be with my brothers in California and to work and see the West and maybe find myself. Now we were on the way home. In September I would enter the University of Missouri at Columbia in the College of Agriculture, where I would learn about animal husbandry; maybe one day I would run a stock farm. That was the message that had been printed under my picture in the Carthaginian yearbook of 1923.

A coyote yelped down near the lake. I raised up on my elbow and looked in that direction; as I did, I saw in the light of my fire not more than thirty feet away another coyote, which suddenly turned and ran.

It was April, and we had planned to travel by the southern route, driving my newly purchased twelve-year-old Harley-Davidson motorcycle and sidecar, because the passes on the northern route were still snowed in. I had saved $125 from my winter's work to get a new suit of clothes and return home to Missouri by train. But my brother Leland had decided that he too would like to go home, only he was without savings. A car seemed beyond our financial means. Driving around town in a Model-T Ford delivery truck with San Francisco Laundry emblazoned on both sides, I had noticed a motorcycle shop of considerable size, which attracted my attention because its sign read Perkins Motorcycles. Leland and I thought a cycle might be about our speed, so we went in and asked for Mr. Perkins. I introduced myself and my brother and explained that I was in the market for a used motorcycle and sidecar, as both of us wanted to go home to Missouri. Perkins showed us several—all out of my price range; I leveled with him, and told him I had $100 to spend. He quickly came up with a 1912 model and an old sidecar to fit. He told us the cycle could stand a little work, but he thought it would do the job for us. I kicked the tires and noticed that all three were bare-faced or nearly so. I assumed a determined expression, and looking him squarely in the eye said, "If you will put new tires on it, Cousin, I'll buy it—but a hundred dollars is all I have." He came right back at me: "Okay, Cousin, you have just bought yourself a Harley-Davidson."

Then, the next day: "Extra, extra. Read all about it. Tourists stranded on Arizona border, running out of food!" Leland on the motorcycle moved to the curb, and we bought a *San Francisco Examiner*. Sure enough, the article told about a growing group of tourists stopped by the Arizona authorities at the border because of an outbreak of hoof-and-mouth disease in cattle in California. A small tent city had sprung up, and there was a shortage of food and water. Many travelers were turning around and going back to California towns.

That was bad news for us. There we were, with the sidecar filled with blankets, pots, other cooking and eating equipment, clothes, and a few other things we just couldn't part with, including a couple of tennis rackets. The sidecar was so packed with gear that the person sitting in it had to stretch his legs straight forward on the pile and hang on with his hands to keep from bouncing off.

After a quick conference, we decided that my remaining twenty-five dollars and the money Leland had in his pocket would not see us through if we had to spend time at the border, so reluctantly we headed up toward Sacramento. We saw the capitol building as we passed by, and it didn't seem to me as attractive as ours in Jefferson City, which was built of white Carthage limestone.

Leaving Sacramento, we speeded up on the paved road and our Harley was running smoothly. We were eating up the miles. Before long we started to climb, leaving the farmland and passing into a more forested mountain slope region. A little later we were in a beautiful evergreen area with a pleasing fragrance. Near the pass there were patches of snow, but the highway was clean. Soon after we passed Lake Tahoe the pavement ended; we started our descent and ran out of the forest into scrub growth areas.

Rounding a curve we saw a cluster of cars and people and some tents. As we got closer we could see a stop sign and men in uniforms. When we stopped, the officials greeted us with instructions to remove all personal goods from the sidecar and take them to tent number 1, where they would be placed in quarantine for an hour. "Drive your motorcycle through the trough, then proceed to tent number two, where you will take off all your clothes and be given a towel. Then you will remain on benches for an hour; you may take your toothbrush and box of food with you, but all the rest of your possessions must be placed in the fumigating tent." An hour later, with our fumigated gear repacked in the sidecar, we were on our way to Reno.

In Reno we stopped for gasoline and asked directions to the Lincoln Highway. "You want to get to Salt Lake City?" We said we did. "Well, the Lincoln Highway is closed as they are widening and rebuilding it. Now everyone has to go south through Carson City, turn east at the end of Walker Lake, go through Tonopah and then north when you get to Utah. It's not the biggest road in the world, but you can get through."

Being nineteen years old, I shrugged and said, "Well, let's go." We were soon bumping along southward to Carson City, where our maternal grandfather had come two or three times with wagon trains to deliver goods to a store there and to sell to people going on to California.

The road on the west side of Walker Lake was rough and dusty. We were about halfway to the southern end when we broke a spring in the sidecar. Nothing to do but unload all our gear and detach the sidecar. We took off the spring, and Leland, being older, said he would ride on to the little town of Walker at the end of the lake, where there might be a garage where he could get the spring fixed or maybe even get a new one.

I made a sort of camp a short distance off the road, gathered some wood from the scrub growth, found a place where there was some sand in a little draw, spread my blanket there, and started my fire. I opened a can of Vienna sausages and one of beans and had my dinner before darkness fell. I felt only a little uneasy at being alone in unfamiliar country. Mother Laura had given me her lady's pistol, a cute little thing that had no trigger guard and fired a .25-caliber bullet. I had never fired it, and the cartridges were old ones, so I decided to test it out. I loaded it, cocked it, and fired at the trunk of the largest bush ten feet away. The pistol went off with a bang, but I saw the bullet kick up dust about three feet to the right of my target. I tried a couple of times more, but that pistol just didn't shoot straight. I decided I would not try to defend myself with that. So there I was, looking up at the stars and listening to the coyotes yelp down near the lake.

Leland returned about two o'clock the next afternoon with a big grin on his face. He had been lucky. A blacksmith had been able to repair the spring. It was no problem to replace the spring in the sidecar, and before long we were on our way. When we came to the edge of town, Leland pulled to the side of the road and said, "And we even have a place to stay tonight, and it's free!"

"What kind of a place?" I asked.

"I'll show you."

He drove a couple of blocks, then turned right, and stopped in front of an unpainted, dilapidated house. The top hinge of the door was broken, and the door hung at an awkward, crazy angle. Bottles and trash were all around the yard. The door was open, so we walked into a dusty front room which led into four other rooms. In two of these were cots with rusty springs.

"How did you ever find this place?" I asked. Leland laughed; then he told me that by the time he had found the blacksmith the day before it was late and the man did not want to start on the spring until morning. Leland asked him if he knew of a place where we could spend the night. "Well," the blacksmith said, "there's an abandoned whorehouse out on the edge of town. Nobody would care if you slept there." He told Leland how to get to the house, so my brother had spent last night there. I'd never been in a whorehouse, but this place wasn't any different from other abandoned houses I'd seen. All the same, I had a strange, guilty feeling, as if I were doing something I shouldn't when I spread my blanket on a cot's rusty springs and lay down.

On we went the next day, and the road got worse and worse. Much of it was made with crushed pumice stone, and before we got to Tonopah we had two flat tires, one of which was beyond repair. We had to buy a new tire in Tonopah, and that took the rest of our cash, all but a nickel. We stopped at a filling station, and Leland was able to give the owner a hard-luck story and get staked to a tank of gas— two and a half gallons. Across the street was a bakery. We could see the baker moving trays of bread to the counter, so in we went, laid the nickel on the counter and said, "Give us the largest amount of bread you can for that."

"You guys look hard up. Tell you what I can do. There's a tray of rolls in the back that got burned on the bottom. You can have them for a nickel."

The baker was already on his way to get the tray, and when we nodded, he put all the rolls into a paper bag. At least we had something to eat.

By the time we got to Salt Lake City the sun was about to set. What a relief to be on pavement again! I leaned over to Leland and said, "I've been holding out on you. I've got fifty cents." What a happy look came over his face! Soon afterward we saw a small hotel, and pulling up, Leland asked for the fifty cents and walked across the street. In ten minutes, back he came and handed me a key to room 202. "While I arrange about the motorcycle, you walk into the hotel as though you were me. The guy will be in his chair behind the counter. Act tired." Not at all hard to do. "Don't stop; just say, 'Hi, Eddie! Boy, what a hard day I've had!' Keep walking to the stairs, turn right, come back nearly to this street, and there's our room. I'll be along as soon as I find some place I can leave the Harley."

After about thirty minutes, Leland came in our unlocked door, and we both slept in our room for fifty cents. The only rub was that we had to get up early to retrieve our Harley from the garage. The night watchman had taken pity on Leland and allowed him to park there free, but the watchman went home at 6:00 A.M. and didn't want to get in trouble with his boss.

We drove down a tree-lined residential street looking for a rooming house. Finally, there it was—a nice two-story wooden house with a sign in the window, Room for Rent. A nice, plump Irish lady let us in and showed us the room. It was just fine and we were able to rent it by the week, room and board, payable at the end of each week. And we could park our Harley in the backyard.

Leland had a white shirt and a tie with him. Because of his good appearance he was able to get a job as a soda jerk in a drugstore. I somehow found out about a place that repacked fresh fruit. It was near the railroad yards. I walked in the front door, spotted the man who looked like the boss, walked up to him, and with a happy smile said, "Hi. Here I am fresh in from California."

"Well, what took you so long," he replied.

"Oh, I had to go clear down through Tonopah, as the Lincoln Highway is closed. Long trip and tire trouble."

"Come along, I'll show you where you work." He led the way to a narrow room with a bench on each side and men and women standing up to boxes of oranges and grapefruit. "This is your spot," he said, and left.

I looked around to watch the operation, and a nice woman leaned over and in a low voice told me to unwrap each grapefruit, inspect it, and if it was bad to place it in a garbage box. The rest were to be rewrapped and arranged in rows in a box, three rows of five to a row, label side of tissue paper up. The job seemed easy and it was, but I wasn't as fast as she was.

After about two hours the boss came in with a funny look on his face. He motioned me to follow him, and when we arrived in the front room he said, "Say, what's your name?"

"Perkins," I replied.

He studied me for a minute and then said, "The fellow I was expecting, an experienced fruit packer, has just come in from California. He too had to go through Tonopah. Sorry, kid, but I had already promised the job to him. I can keep you for the rest of the day, however, if you want to do another job for me. Follow me and I'll show you what it is." He led me to a boxcar at the other end of the building and explained that my job was to clean out the boxcar, put all the spoiled onions in garbage cans, and then sweep the floor of the car. I had a shovel and a broom, and despite the unpleasant odor of rotten onions I went to work. At five thirty the boss came back, okayed my work, and handed me three dollars. At least we had spending money.

Next day I got a job at the Utah Oil Refinery. It was a laboring job, separating bricks and chipping off the mortar from an old building. Sometimes, putting on hip boots, I'd help clean out a huge oil tank, entering through a manhole and with a big squeegee pushing the sludge left on the bottom to a drain in the middle. Six to eight of us worked at a time, and we were limited to six minutes in the tank. The gas from the sludge was terrible and couldn't be breathed any longer than that. I was in and out four or five times and I had a splitting headache.

After a month with both Leland and me working, we figured we had enough saved to see us the rest of the way home. I told my boss I had to quit and asked when I could get my pay. He explained that payday was every two weeks. I said I couldn't wait that long; I'd just had word my mother was ill and I had to go home as quickly as possible. He saw through that one, and tried to get me to stay on the job. I explained I couldn't stay on, and asked again how I could get my pay before the end of two weeks. The only way, he explained, was to get fired. So I said, "All right, fire me."

"Oh, I can't do that, kid. I've had my eye on you and you could go far in this company. Besides, if I fire you it will be a blot on your record that will follow you wherever you go."

I pleaded a bit longer, but he refused to fire me.

The next day I showed up for work at eleven o'clock. The boss was mad at me and said, "Where you been?"

"Oh, I didn't feel like coming to work on time today."

He said something uncomplimentary in a growl in the back of his throat, and added, "Come here, I've a job for you at the coal pile."

When we got there, he said, "Here is the hose. Start watering the coal." The pile was about thirty feet high and a hundred feet long. As soon as the boss left, I sat down next to a post; I didn't even turn on the hose. In about an hour the boss came back, saw the situation, and said, "Oh, come on. I'll get you your pay."

Our route from Salt Lake City took us east up a long canyon that became more and more narrow as we climbed. On a switchback we could look down on all of Salt Lake City and away beyond to the lake. Up and up we went, noting the changing vegetation and cooler air. Clouds began moving in, and about the time we reached the summit it was snowing hard. Our Harley did not have the traction of an automobile, and as the snow piled up on the road we were not only slowed down, but frequently were stuck in the ruts left by the cars. The wheels of our sidecar were somewhat short of the right rut and that made driving more difficult. One of us rode the Harley, and the other lifted and pushed. The snow was by now knee deep, and we were having a terrible time. In late afternoon, wet and tired to the point of exhaustion, we decided to stop where we were for another night. We had some food with us, and blankets. We heaved the Harley off the road and made camp in a grove of evergreen trees. We scooped away the snow, broke green branches and piled them up in a sort of bed to keep us off the ground, gathered wood, and made a fire. Our blankets were more or less dry, and it felt good to wrap up in them and lie down. Before going to sleep I looked past the fire and saw three wolves looking at us. I moved to see better, and they faded into the night. "Good night, Brother. Some Memorial Day!"

It took us three days to go forty miles. "Driving" consisted of pushing the Harley and trying to get it started; then the man next to the cycle would jump on and the machine would move a few feet before getting stuck again in the deep snow. The other man would catch up, and we would try again. Passing cars offered encouragement, but could not give us any real help.

At last, the snow became shallower, and eventually we were out of it and on a wet gravel road. The sun came out as we drew up to a filling station at Rock Springs, Wyoming. The road ahead was a wide gravel highway; we were dry and warmer and decided to push on. It was Leland's turn to drive, so I relaxed as much as I could in the sidecar. About ten miles out of town the road became a washboard, and the Harley was bouncing and vibrating from one bump to the next. All of a sudden we heard a sharp snap and skidded to an abrupt stop with a broken front fork. We pushed the Harley and sidecar to the edge of the road and sat down on the low bank in despair. A cloud of dust in the distance turned into a small truck. As it came closer it slowed and then stopped, and a friendly young man asked if we needed help. Boy, did we need help! We explained about our broken fork; he got out to look at it. Then he said, "I'm going to Rock Springs. Maybe we can tie the front fork that isn't broken on my tailgate and I can tow you in." He produced a rope and tied the Harley to his truck so that the front was off the ground but the other two wheels were not. The arrangement worked just fine.

Leland rode in the cab while I stayed in the back end to keep an eye on how the Harley was trailing. Talking to our newfound friend, Leland discovered that he knew a young Italian man in Rock Springs who not only rode a motorcycle but repaired them. He drove us to the cyclist's house. The woman there said her brother was not at home, but told us where he was. Our new friend suggested we untie the Harley and leave it beside the white picket fence. He then drove Leland to find the cyclist. I stayed with the Harley, sitting in the sidecar in the warm sunshine.

After a while a dark-haired young woman carrying a baby in her arms came out of the house and said hello. She asked where we were from and where we were going. During a lull in our pleasant conversation, she came closer to the fence and said, "Are you all right?"

"Oh, sure," I replied.

She said, "Have you got fever?"

"I don't think so," I said.

She reached across the fence and put her hand on my forehead. "You sure do. Tell you what. I think you are sick. Where will you stay tonight?"

"I don't know yet; I'll wait for my brother to come back and then we can find a place."

"When my husband gets home from work in about an hour, I'll ask him if you can stay here with us. We can make a place for you and your brother." She turned and went into the house.

In about an hour Leland returned, with a tall fellow carrying a motorcycle fork. I was shaking with a chill, but I was glad to see that fork. Leland explained that Joe here had remembered seeing a frame of a 1912 Harley lying out on the town dump. They'd gone there and taken the fork off, and sure enough it fit our Harley.

About that time the nice lady came out of the house with her husband. There was much talking in Italian, and he came over and felt my forehead and nodded his head. Joe and Leland felt my forehead too and agreed we should accept the invitation to stay overnight.

My bed was on a screened back porch and dining area, and directly over my cot hung a cage for breeding canaries. By the time I got to bed I could feel my own fever, and I knew I had the flu or pneumonia. I took the prescribed aspirin and other medications and stayed in bed. After a week I began to feel better. On a Saturday evening after dinner, the husband spoke with his wife, who interpreted to us that he knew how to make me well quickly. He took a glass pitcher to the chicken house at the bottom of the deep lot and returned with the pitcher full of red wine. Next to it on the linoleum-covered table he placed three tall glasses, filled them with wine, and passed one to Leland and one to me. He said, *"Salute!"* and indicated a toast. We sipped our wine. He took a big gulp. He talked to us in Italian, and my high-school Spanish allowed me to catch a word now and again. But it was not a stimulating philosophical discussion. Every once in a while a child would cry, and our host would go into the house and silence it. As soon as he left, Leland and I would pour our wine back into the pitcher. When he returned, he would refill our glasses and indicate that we should drink up. The second time he left we noticed he was a little unsteady on his feet. I had drunk about a half a glass of wine and I was feeling it. When he returned, I indicated in Spanish and by pantomime that I didn't feel well and should go to bed. He and Leland drank for a time and then they turned in too.

The next morning I awoke feeling just fine. That Italian wine did indeed have medicinal qualities.

Next day we repacked the sidecar, but left the tennis rackets and some other goodies for our hosts. When we got home we sent a box of presents for each member of that wonderful Italian family.

In Denver we called our cousin, Carson Perkins, who was a vice president of a bank, but he was sick in bed so we did not get to see him.

The journey through Kansas was uneventful. By the time we got near the eastern part of the state the alfalfa hay was being cut, and as we drove by the fields the air was filled with its fragrance. Our homecoming was emotionally charged. As I kissed Mother Laura, she cried; and so did I.

4

Snakes Are
Good for Farmers

That September I went to the University of Missouri at Columbia. Its well-known College of Agriculture offered courses in zoology and animal husbandry, and I had decided to enroll in that school. It was during this period that my interest in conservation started. Every year the College of Agriculture held an agricultural fair in which all members of the school were asked to participate. I suggested arranging an educational exhibit to show farmers the value of the snakes on their farms. I hoped to discourage the usual practice of snake killing. The project was agreed on, and I wrote a little pamphlet to be given away to visiting farmers. In it I pointed out the importance of snakes to agriculture, and emphasized that many of the snakes found on farms feed on mice, rats, gophers, and ground squirrels, all of which are detrimental to the farmer because they compete with him for his grain. I worked out the value of a snake on a farm in that part of Missouri at that time in terms of the quantity of grain that would be eaten by a mouse and the progeny of that mouse over a year's time. The snake would have consumed enough mice to have saved the farmer about five dollars in grain.

I had enrolled in the College of Agriculture because I thought my interest in animals lay in the field of animal husbandry. That first year, therefore, I took an animal husbandry course in which I expected to learn how to run a stock farm and how to raise and breed various domestic livestock. Unfortunately, the course taught none of that, but concentrated on judging various breeds of cattle, sheep, swine, and horses. It was long, too, on statistics that compared the numbers of cattle, swine, and sheep received at various stockyards around the country.

At the same time I was learning in a general zoology course about wild animals. Before the year was finished I knew I must leave the College of Agriculture and enter the College of Arts and Sciences. After making this transfer, in my second year I took a course in evolution under Dr. Curtis, head of the department of zoology, as well as courses in comparative anatomy, botany, chemistry, and geology. At the same time I read all the books in the library on snakes and other reptiles, and before that year was out I knew I wanted to work with wild animals.

If I was not prepared for the loss of my mother, I was even less prepared for the sudden loss of my beloved stepmother. One evening during my second year at the University of Missouri I received a long-distance phone call from Father telling me that Mother Laura was ill and suggesting that I come home as soon as possible. I caught the train the next day, and late in the afternoon arrived in Carthage. A friend, Everet Blosser, met me at the station and drove me straight to the hospital. I could tell by his silence that Mother Laura's illness was more serious than I had realized, and when I dashed into the hospital and was taken to her bedside, my tears started to flow before I could say a word. She took my hand and started to speak, but the deep, hacking cough of her pneumonia was too much for her. My grief was too much for me; and I couldn't tell her any of those loving things I wanted to say. I knew she was going to die, and she did that night.

In the spring of my second year at Missouri I happened to ask a senior I knew from a geology class—he was an assistant in the lab—what he was going to do when he graduated. I was thinking in terms of his geology major and wondered if he might be heading toward a job with an oil company or in mining. I wasn't prepared when he replied, "Oh, I think I'll go back and work in my father's store."

This was a shocker. He had spent four years studying geology, and yet he planned to forsake that and help his father in the store. This made me wonder how many others graduating were also planning to work in some field other than the one they had prepared for. I thought I had better find out, so I asked a number of seniors, and more of them were going into the stock market or planning to sell automobiles or real estate than were going to work in the fields of their academic majors.

I asked myself, "Am I going to spend another two years studying zoology, hoping I can work with wild animals, and then wind up in some other field?" It seemed wasteful to spend my retired father's money for something I might never use. Was I sure I wanted to spend my working life with wild animals? I thought so now, but perhaps I should find out as soon as possible.

During spring vacation I went home to Carthage and had a long talk with Father. I laid the whole question out before him. I told him I felt that to spend four years preparing to enter a specific field and then, when the time came, to switch around and do something quite different, didn't make good sense. It seemed to me that I should leave school and try for a job at the Saint Louis Zoo, which I knew had a fine reputation. If I could get a job, I should be able to determine whether working with wild animals was the right field for me. If, after a good trial period, I found it was not, then I would try something else.

At this my father said, "Marlin, I agree with you. Perhaps wild animals will be your field, but after you give it a good try, if you decide you would rather do something else, then you can go back and finish your education." Having spent so much of his time as a circuit-court judge helping boys brought before him at the bench with their problems, Father had the understanding to help his third son with his own perplexing problem. Bless him for being so knowledgeable and so tolerant.

5

Life with Mack and Blondie

I supposed that such a glamorous institution as the Saint Louis Zoo must be inundated with job applications and that if I applied in person I would run less risk of being turned down or lost sight of in the pile of letters. As soon as classes at the university ended in June of my sophomore year, I drove to Saint Louis with Leland in his Locomobile. My brother had decided to seek a job in the big city too. Once we were established there in a modest rooming house, I proceeded to the zoo, walked into the new monkey house, and asked the keeper on the floor where to find the head office. He pointed to the stairs at the north side of the building and suggested I ask for the director, George Vierheller.

I found Vierheller working at a desk in a small office on the second floor. A slim man in his forties, with keen blue eyes, receding brown hair, and a serious but friendly manner, he invited me to sit down on one of the plain wooden chairs with which the office was furnished. I explained that I hoped to make a career of working with wild animals and wanted to apply for any position open at the zoo. I pointed out my long-lasting interest in animals and my emphasis on conservation and showed him the pamphlet I had written for the farmers' fair at the University of Missouri.

"Well, young man," he said finally, "if you really would like to work with us here, I do have a job for you." That was a great moment for me. "You report tomorrow morning at eight o'clock," Vierheller continued, "at the back end of this building, to Barney Meyer. He's in charge of the grounds and maintenance. There's hard work to be done, but if you're interested I can start you there, and I'll be happy to have you on the payroll. Your starting salary is $3.75 a day, six days a week."

I thanked him and told him I'd certainly be there.

My first job was sweeping sidewalks. My second was trimming the hedges in front of the new bear pits. I also mowed grass and did other odd jobs, such as digging ditches. In about two weeks Vierheller came by and said, "I have a job up at the east side of the grounds for you. I'd like you to work as a relief animal keeper. Come along, I'll get you started."

I was introduced to Floyd Smith, an animal keeper. "I'd like to have you work with Floyd the rest of the day," Vierheller said. "He'll tell you what to do."

"Come on over here, kid," Floyd said, "we have a job to do in the yard with the Rocky Mountain goats, sweeping and raking. But before you go in there, let me caution you. As long as you are going to work with animals, you must realize that all animals can be dangerous. Why, even a mouse can bite you, and that can hurt like the mischief and maybe even become infected and cause the loss of a finger. So be careful, kid. Watch the animals. Don't get too close to them, and don't make any quick moves."

I was in heaven. These were the first Rocky Mountain goats I'd ever seen. Such stately, beautiful animals, and so white. I noticed their very sharp, black horns, and I could understand why Floyd had warned me. In the center of the fenced yard was a nice building where the three animals could sleep; there was a separate compartment for their food. All three stayed close together, and I noticed that the male was considerably larger than the females. Next to the building was a steep wooden ramp that ran diagonally up the back, somewhat past the roof, with two jogs that took it up to a high platform where the animals could climb and look over that part of the zoo. The yard to the right had a similar arrangement for bighorn sheep. To the left were zebras, and across the visitors' path were the ostriches, at the end of a wooden structure called the ostrich house. Facing this, just across another visitors' path, was the lion house.

A few days later, I was told to rake the yard of the South African white-tailed gnu or wildebeest. There was only one animal in the yard and as I walked slowly in to get the lay of the land and see where to start, the gnu suddenly advanced toward me. I realized he wasn't kidding; he was really threatening me. I was afraid to hit him with anything, particularly the rake I had in my hand, so I just stood there. He danced back and forth, menacing me with his horns, until one of the keepers threw me a broom. I reached down, picked it up, and stood still, wondering whether I should hit such a fine animal. One of the boys called, "Hit him, hit him," so I made a pass at him and missed and hit the ground. But the threat on my part was apparently enough to make the gnu change his mind. He went over to the other side of the yard and left me free to rake.

My new job as relief animal keeper put me in contact with all parts of the zoo. I became acquainted with all the keepers and their animals. Once a week I took a turn in the monkey house. The capuchin monkeys, from South America, soon became my friends. There were also weeping capuchins, monkeys with a happy, yet plaintive, whistlelike call, climbing the scale and ending in a very high tone that I had trouble matching with my own whistle. These two species of monkeys were the types organ grinders in cities almost everywhere used to attract attention and to collect coins.

Another large cage held a family group of spider monkeys, with a dominant male in the alpha position—that is, all the others had to defer to him.

I'd been at the zoo a few weeks and felt I had been accepted by most of the monkeys I took care of. One day I noticed that the wall in the spider monkeys' cage needed scrubbing. The tools were in the keepers' passageway in the rear of the cages. After flushing the cage and wall, I entered via the access door carrying a scrub brush. Mike, the big male, pursed his lips, shook his head, and smacked a friendly greeting. I responded in the same manner and turned to attend to the wall. Mike made a long jump and landed with all four legs and a prehensile tail around the handle of the scrub brush. I held firm to see what his next move would be.

"Mike," I said, "you know I have to scrub that wall, so be a good boy and let me get on with it."

I moved as though to start on the wall, and Mike screamed and let me have it. Quick as a flash, he slit a one-inch gash through the muscle of one of my hands, then turned and climbed to the top of the cage, still screaming and threatening me. I quickly changed my mind about washing the dirty wall and, leaving a trail of blood, went to report to the head monkey keeper, Max Mahl. He got out the first-aid kit and patched me up, telling me over and over, "You can't trust dem monkeys."

Another part of my job in the monkey house was to clean the large cage in the center of the building. This was a forty- by twenty- by forty-foot wire cage with oval ends, surrounded by tropical foliage and lighted by a skylight over the whole section. The routine was to feed the rhesus monkeys first thing in the morning. They were given two buckets of chopped carrots, sweet potatoes, lettuce, oranges, apples, bananas, cabbage, and whole-wheat bread. After they had finished their breakfast, I would flush out the cage with a stream of water and enter with a scrub brush and an empty bucket. The pool in the cage was drained, and I started scrubbing, keeping an eye on those active monkeys, who were presided over by another monkey, also called Mike.

Rhesus macaque monkeys come from India. There they live in large troops, some near temples, where the monkeys help themselves to rice and other food offerings to the gods. I decided my best ploy was to be so busy, scrubbing the pool and the concrete floor of the cage with such vigor, that the monkeys would stay away from me. I became so interested in the exercise I was getting that I doubled my efforts. I hadn't played tennis in a while and I needed the workout.

I never lost sight of the monkeys sitting and playing in the several trees overhead, particularly Mike. One day, as I was leaving the cage, Mike came down, smacked lips in friendship, and turned around and presented. This is a social activity in these animals, and it usually suggests not sex, but simply friendliness. Imagine my surprise when I saw that Mike was a female. She was the largest monkey in the troop and the alpha animal. Someone had given her to the zoo, saying she was a male.

Within about two months after joining the zoo staff, I was placed in charge of the reptiles. There were not very many, and they were not on exhibit, but were kept in the basement of one of the buildings. They included a fine boa constrictor, some big bull snakes, an indigo snake or two, and an assortment of king snakes, spotted black snakes, and a coachwhip. In addition, there were some turtles, both land and water species, and quite a few alligators.

Not long after I began working with the reptiles, Vierheller told me, to my excitement, that he hoped soon to build a real reptile house. He had proposed this plan to the zoo's Board of Control, but they were concerned about the acceptance of such an exhibit by the general public. Vierheller thought the best way to convince them and to let the public know what was going on at the zoo would be to build a temporary exhibit to be displayed for a couple of months that summer. He asked me to give him some designs for cages and to write the educational labels for the exhibit itself. Then he would buy a couple of nice big pythons and install them in a large cage to show what some of the exhibits in the proposed new building would be like.

All this was carried out in quick order. The temporary exhibit occupied a small portion of the old ostrich barn back of the lion house. Vierheller purchased two monster reticulated pythons, and we placed them in their new home—a large exhibition cage with a background painted to show the natural habitat of these animals in Southeast Asia. In the cage was a section of a tree for the pythons to climb on. One animal was dark in color and twenty-two feet long. He was so big and massive that we named him Mack, in honor of the popular truck of that name. The other snake was a nineteen-foot specimen, lighter in color, and since she was a female we named her Blondie.

The small snakes and lizards and Mack and Blondie proved so irresistible that in two weeks' time there was a line of people a block long waiting to pay a Sunday visit to the new snake exhibit at the zoo. Because of the limited space and the huge crowds of people, it became necessary to recruit several zoo employees to keep the single line of people moving. Most visitors wanted to stop and see every reptile on exhibit. To have allowed this would have deprived others of the opportunity to see anything, so we had to repeat, "Keep moving, please. There's a long line behind you and all want to see too."

The members of the Zoological Board of Control noted the enthusiasm with which this temporary exhibit was received by the zoo's visitors and realized that people liked to see snakes and other reptiles. The board soon sanctioned the design and construction of a reptile house.

During the autumn and winter, zoo architect John Wallace was busy designing the new building. Vierheller and I spent long hours with him going over details and offering practical suggestions on cage sizes, door arrangement, drain positions, water supply, and temperatures needed by cold-blooded reptiles. We discussed the lighting of cages and the problems of reflections on glass cage fronts that could block the view of the reptiles inside. A food-preparation room was designed behind the cages, and space was provided in the basement for boxes in which to raise meal worms, dermestid beetles, crickets, earthworms, and fruit flies. A small office was designed in another section behind the cages. This would be the nerve center for the building.

After the completion of the new building in 1927, we set aside a day to move Mack and Blondie into their new quarters. We got out the two big, trunklike boxes they had arrived in. While neither was a vicious snake, we decided we'd catch Mack first. In order to keep Blondie from being overexcited by the movement of several people coming into the cage to lift Mack out, I decided to go in first, catch Blondie, and gently slip a black sack with a drawstring over her head. Then I could turn her loose and she wouldn't see anything and so wouldn't be upset by the activity that was to follow. Everything went well until I had knotted the string around her neck. Then, when I let her go, she pulled back, and as she did so, her head came out of the black bag. She cocked her neck, head pointed straight at me, and I realized she was assuming an aggressive attitude. By this time about four fellows had filtered into the cage to catch Mack. Moody Lentz, who worked with me in the reptile house, already had him by the neck. I said, "Look out, fellows, she's loose! We better get out."

Moody let go of Mack and started pushing the other fellows through the door. But it was a small door, so climbing out took a little time. I kept analyzing the situation and watching the movements of the snakes; I moved to the other side of the tree in the cage. Then big Mack got excited and started crawling forward with his neck pulled back in an S-shaped loop, ready for the strike. He came so far forward that my escape was cut off. I flattened myself into the rounded corner of the cage. Mack struck just as Moody slipped out the door. He struck so hard and so far that his head and about a third of his body were outside the cage. Luckily he missed Moody, who picked up a broom and tried to force the snake back into the cage. But by that time Mack was so excited and angry that he started after Moody, striking as he came. Moody backed until finally he had to step up on some feed boxes and then grab hold of a roof beam above him to raise his feet up, out of striking range.

Inside the cage, I heard all the commotion, the shouts and warnings of several people. Blondie was still in the right-hand corner of the cage, but cocked and ready to strike. I began inching my way toward the door and when I thought I could make it I bounded out. Blondie struck, but she didn't come out of the cage itself, and I was able to close the door.

I then saw that only Moody, big Mack, and I were left in that corridor of the building. Some of the other fellows had taken refuge in stalls, around in front, and two were way up on top of the building rafters. One fellow had pulled the doorknob off trying to get outside and was quivering in a corner. I joined Moody on top of the feed boxes and handed him a gunny sack. We were directly above Mack, who by now had quieted down a little.

"I'm going to drop that bag right on top of his head," Moody said, "and drop down alongside him at the same moment."

"Okay," I said, "I'll follow you." Moody had hold of big Mack about the same time the bag touched the python's head and neck, and I dropped down and backed Moody up about a foot or two below his hands. Seeing this, the other fellows came back and grabbed the big snake; we had no trouble then putting him into the trunk. We waited until the next day to catch Blondie so she would have a chance to quiet down. We had no special difficulty getting her into the trunk and moving her to her new accommodations.

Mack started to feed just fine. He was a wonderful snake and calmed down nicely in his new quarters in the reptile house. Blondie lived in another cage there—a glass-front cage with a mural painted by Frank Nuederscher depicting her natural home in India. The cage was fourteen feet across the front, eight feet deep, and about ten feet high. It contained a nice pool of water, big enough for her to get her whole body in, and a tree for her to climb. There were some rocks fitted into the round corner of the cage.

Blondie must have had some unknown food preference, because she never ate of her own accord. After nine months of offering her every kind of food we could think of, we realized she was losing weight and we had better do something about it. I got a six-foot section of rubber pumping hose, two and a half inches in diameter on the inside. I smoothed and rounded off one end so it wouldn't scratch Blondie's throat, and made a plunger of rubber attached to a long metal rod. This was to force the meat through the tube into Blondie's throat.

Five or six freshly dressed domestic rabbits were ground up, bones and all. To this was added some ground horse meat to make fourteen and a half pounds. The meat was fortified with vitamins and trace minerals. We lubricated the inside of the tube with slippery elm and filled it with this rabbitburger.

We were ready for the operation of force feeding Blondie. But first we had to catch her. I recruited six strong young keepers from throughout the zoo, and at the appointed time of one thirty, we were ready to begin the operation.

The first thing to do was to get hold of Blondie by the neck before she could get hold of us. Luckily, the cage had two doors, one in the back and the other in the side. Moody Lentz was number-one man. His job was to grab the snake by the neck. He opened the back door slowly and noiselessly, and stepped gently up into the cage with a gunnysack in his hand. Then he took a couple of steps very slowly, in a crouched position, toward the snake, who was lying coiled up in the corner of her cage, near the glass. Moody watched her closely to see if there was any sign of movement or awareness that he had entered the cage. Seeing none, he got into the correct position, tossed the gunnysack over her body and, at about the same time it landed on her, grabbed through the sack to get hold of her neck with both hands. He then got a good, firm grip with one hand from underneath and freed himself of the sack, which was left in the cage. I came in behind Moody as soon as he had hold of Blondie's neck. I backed him up a foot or so down the python's body. The number-three man was behind me, followed by numbers four, five, six, and seven. Then came number eight, the man holding Blondie's tail and pulling backward to keep her stretched out. By this time Moody and I and the others behind me were making our way across the cage and out the side door onto the floor of the building with the snake securely held in our arms. She, of course, was twisting and squirming as much as she could, but Moody had a firm grip on the business end of that snake, and nobody was going to be bitten by her.

When everything was in order and the men had sat down with the snake across their laps, each with a firm hold of his section of her body, Moody assumed a kneeling position and I came up slowly toward the python's head with a pair of long forceps. I slid these gently in from the side of her mouth and then let them spring up, opening her mouth. I shifted position to get one finger in the front of her upper jaw and a finger of my other hand in the front of her lower jaw. Then, with gentle pressure, I opened her mouth until it was about 180 degrees open. We were now ready for the insertion of the tube, which had by this time been lubricated on the outside too. A keeper named Jake Schoenberg was in charge of this phase of the operation. He gently placed in Blondie's mouth the end of the tube, which she bit. I disengaged her jaws again, and with a gentle pressure Jake slowly forced the tube down her throat about a foot. Changing position I began massaging the food toward her stomach, which was about in the middle of her body. She was so big around that my hands soon grew tired, as the meat was being forced down between the ribs, which lay on both sides of her body. Carefully, I used my knee to spread the ribs apart and ease the food down and she helped by forward-crawling motions. This was a gentle operation. We didn't want to bruise or injure the snake.

While I was doing this, Jake was refilling the tube for a repeat operation. After massaging the meat down into her stomach the second time, I examined the snake for any parasites and lifted out with my forceps all pieces of bone or meat that were left in her mouth. Then, with a mild antiseptic I cleansed her mouth to prevent infection as a result of the force feeding.

We were now ready to put Blondie back into her cage. This called for a little preplanning. Here were eight men holding a nineteen-foot python whose body was as big around as a small tree, and we had to get her back into the cage uninjured without any of us being injured either. Moody took position by the back door, and I put my knees next to the wall of the cage, ready for my action. Moody stepped into the cage and started for the other door. I fed my section of the snake into the cage and then the next section from the fellow behind me, and so on down the full length of the snake, until Moody stepped out the side door, holding the snake's head, while I, crouched at the other door, had her by the tail. We exchanged okays; then each of us tossed our respective ends of the snake into the cage and slammed the doors shut.

We followed pretty much the same procedure each time it was

necessary to feed Blondie. Usually, on first being returned to her cage, she would crawl quickly around the cage or take a defensive position with her neck coiled in an S-shaped loop, ready for action. In this mood she would strike at any moving object. For that reason, we always roped off the section in front of her cage so that visitors could not come close enough for her to see them until she had quieted down, usually after forty-five minutes to an hour. Then she would curl up in the corner of her cage and sleep for several days, digesting her food.

One cold, blustery winter day, with a blizzard blowing outside, was force-feeding day for Blondie. Forty-five minutes after she had been put back into her cage, I came out from my office to see if she had quieted down enough so that we could take the people-barrier ropes away. There were only two visitors in the building, and as I walked up to the ropes one was saying, "Man, man, isn't that a big old snake!"

"Ah, go on," the other one said. "That isn't any real-live snake. I saw a bunch of men here pumping it up with a big pump awhile ago."

Blondie's force feeding became a regular feature at the Saint Louis Zoo. The press would publish advance notices of her feeding day. *Life* magazine covered the operation, and many other publications as well. When the weather was good, we frequently had 3,000 people on hand to watch the feeding. We continued to try to get Blondie to eat voluntarily, but she never would. She did, however, set a world's record by living twenty years at the Saint Louis Zoo, all that time being force-fed. So far as I know, that record still stands.

Mynta Mae Miller Perkins,
my mother

Judge Joseph Dudley Perkins,
my father

Me at four

Leland, Carlyle, and I with our billy goat and cart

This picture was taken while I was a student at the University of Missouri. (*Photo by Parsons*)

As a young curator of reptiles at the Saint Louis Zoo, with a tame python.

(OPPOSITE) Moody Lentz and I, with our snake sticks, hun snakes in a Louisiana swamp. (*Marlin Perkins Colle*

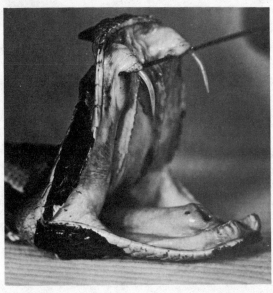

The gaboon viper from Africa that bit me on the last day of 1928. This viper has the longest fangs of any snake—up to three inches in length.

Force-feeding Blondie, the nineteen-foot reticulated python before three thousand spectators at the Saint Louis Zoo. (I am the one operating the pump.)

(BELOW) Moody Lentz, Pete Miles, and I moving a giant komodo monitor Lizard, brought by Pete from Indonesia, into an exhibition cage at the Saint Louis Zoo.

George Vierheller and I
in 1931, returning to
Sarasota, Florida, from
snake hunting nearby.

(BELOW) At left, I am
holding a sumebadora
snake, while Moody Lentz
holds a culebra mico
snake, both from our
Central American
expedition of 1930.

(ABOVE) Displaying a red
t snake at WBKB
perimental TV station
Chicago, 1946.

aring a joke with my
end, the chimpanzee
einie II. (*Photo by Gates
est, Chicago Park District*)

The baby gorilla Sinbad on his arrival at Lincoln Park. He weighed
eleven pounds and needed and was given a lot of tender, loving care.
(*Photo by Gates Priest, Chicago Park District*)

(LEFT) Bushman, the gorilla, was our most popular animal at the
Lincoln Park Zoo. He reached a height of six feet two inches and a
weight of 570 pounds. (*Photo by Gates Priest, Chicago Park District*)

(LEFT) Incising the bite of the timber rattlesnake that bit me just twenty minutes before air time for "Zooparade" on April 1, 1951.

(RIGHT) Applying suction to the incised finger to remove as much venom as possible.

(BELOW) Analyzing the unfortunate bite, I decided to go to the hospital. Lear Grimmer, right, had to host the show that day and for the following weeks. (*Photos by Gates Priest, Chicago Park District*)

6

Electric Eels and Viper Venom

During the Depression, banks were closing around the United States, and it happened that I had my money in a small bank on Delmar Boulevard, which also closed. Not wishing to trust another bank for a while, Moody Lentz and I decided to keep our valuables in a very secure place. A series of North American habitat cages had been developed for one quarter of the reptile house. One of these was to be the cage for copperheads. The habitat display showed a scene in the Missouri Ozarks. To make this as realistic as possible, some weather-etched limestone rocks were set into the cage. They were continued into the mural painting to look like an outcropping on a hillside. Next to the rocks was a little pocket meant to be used to hold native plants, and we usually had a young Missouri cedar tree growing there. The cage contained a pool of water, and the floor bordering the pool was covered with tanbark and peat moss.

The two flat stones on that little ledge of rock lifted off, and there was a hollow place underneath. This was our safe for our valuables. Not only money, but our watches and even our house keys were kept in there. With the cage safely locked and nine or ten copperheads in residence, we didn't feel anybody would disturb our safe, even if they knew where it was.

In a much larger cage next to the copperhead quarters was a painted background and floor designed to represent a scene in the semiarid southwestern part of the United States. Within this, we kept diamond-backed rattlesnakes, southwestern coachwhip snakes, and some bull snakes. Occasionally we would add such creatures as chuckwalla lizards, which live in the same region. Chuckwalla lizards are largely fruit eaters. We found that they liked, best of all, yellow flowers and bananas. In order to get them to feed during the winter months, we had to coax them a little. Carrying a banana, I would go to the back door of the cage, open it cautiously, and look around to see where the rattlesnakes were. If they were not near the door, I would offer little pieces of banana on the tip of my finger to the chuckwalla lizards. One day when I was doing this, they were feeding quite well, and I slowly sat down in the door of the cage facing the low ledge where the lizards were. I kept breaking off pieces of banana and offering them first to one lizard and then to another, and I was so intent on feeding the chuckwallas that I forgot to monitor the rattlesnakes. My first indication that a rattler was near me was a movement I caught out of the corner of my eye. I turned my head just enough so I could look down, and there was the head of the biggest rattlesnake in the cage, about six and a half feet long. It was well within striking distance! I realized I didn't have time to get out of the cage, so I just lowered my left hand on to the rocky ledge and remained as still as possible. The snake came to my right arm, touched it with his tongue, crawled across it into my lap, his tongue playing all the time, then across to my left hand, and onto the ledge itself. Finally all six and a half feet of him crawled across my lap onto that ledge. Only then did I feel safe enough to retreat from the cage. The rattler was only curious; he wanted to know what was going on, and as long as he wasn't threatened, I wasn't either.

I've come to learn that each animal has its own temperament, just as people do. Some appear to be naturally aggressive and quick to anger; others are gentle and slow to become aroused.

On the other side of the copperhead cage, in another habitat setting, was a display of timber rattlesnakes. One individual in this cage had been captured by Moody and me at an abandoned sawmill, now overgrown with blackberry briars, on the old Schexneider Plantation up the highway from Gramercy, Louisiana. He was nearly six feet long and big around and as treacherous and mean as any snake I've ever seen. We had to be careful every time we approached that cage to clean or to feed. When draining the pool, we had to keep a sharp eye on that snake because he would slowly creep toward us with his neck arched in an S-shaped loop trying to get close enough to strike. The strike of a snake is simply the straightening out of its body from that S-shaped loop, which gives a longer reach. And a strike comes so fast from a coiled snake that the target has no chance to dodge out of the way. This timber rattler lived for a number of years, and never once did he quiet down to a point where you could trust him not to attempt a sneak attack. I feel he represented an extreme example of an animal going out of its way to attack. Most animals are willing to assume a passive attitude, a live-and-let-live spirit of coexistence, although anyone working with animals must be aware that almost any animal may attack. Most of these attacks, from the animal's viewpoint, constitute self-defense.

The cage on the other side of the timber rattlesnakes' quarters was for cottonmouth water moccasins. This was painted to resemble a cypress swamp of the South, with real cypress knees implanted in the cage floor near the spacious water pool. For a time, we had seven cottonmouths in that cage, and we had seven personalities among those snakes. At one extreme was a moccasin very similar to the canebrake rattlesnake next door. He could not be trusted at all. At the other extreme was a cottonmouth so gentle it actually could be hand fed if you held up a herring by the tail and let the snake reach up and slowly swallow the fish, starting at the head end. Moody and I always felt that we could really have picked that moccasin up and carried him around like a nonpoisonous snake. But, of course, we never did.

One day two of the moccasins in this cage put on a performance we had never seen before. They crawled toward each other, and as their noses touched, they elevated their heads until finally these three- or four-foot snakes had half their bodies in the air, writhing with a movement similar to Indian wrestling. The neck of one was pushed against the neck of the other, and it was obvious that much pressure was being applied to force the other snake down or away. On occasion, one snake would move and slip, and the other would be flipped halfway across the cage. At first we thought this was a premating performance, but it wasn't long before this activity was described by another herpetologist as a struggle for dominance between males.

When I realized I would like to stay at the zoo, I decided I should acquire still more education. I enrolled in night school at Washington University and for several years took courses in advanced zoology, Latin, and psychology. In addition, I read books on herpetology, mammalogy, and ornithology. I became a member of several scientific societies and received, read, and filed their publications. I started building my own library for my work as a zoo man.

This self-education process has continued all the rest of my life. Besides my reference library, so important to me in my zoo work, I have accumulated volumes on archeology and anthropology, travel, and many other subjects. To this day, my reading is more likely to be informative than passive. When traveling I usually carry such magazines as *Time, Newsweek, The Reader's Digest, Modern Photography, Popular Photography, Oceanus, Natural History,* the Audubon Society magazine, *Science News, Science,* and *Omni.*

The more I learned about animals, and the better I understood ecology—the relationship of an animal or plant to its environment— the more I realized the imbalance of habitats caused by man when he plowed the prairie, felled the forest, or drained the marsh. The plants of each area were largely eliminated and most of the animals dependent on them disappeared as well. The prairie grasses and forested slopes were replaced by farms; marshes were dried up; and sometimes the land that was left proved unsuitable for agriculture.

In the early days of the United States the feeling was that this did not matter, for there was always another place further west to claim for man's uses. The land beyond seemed an unlimited Eden to be mastered. But as our population increased, the land began to fill up, and less and less of unmolested wilderness remained.

We have become the most powerful nation on earth because of our seemingly unlimited resources. We have used them lavishly for the industries of our free-enterprise system, of which I am so proud. But when our resources began to become scarcer, we had to barter with the rest of the world for the raw materials needed to support our ever-developing economy.

In our headlong dash to maintain our economic superiority over the rest of the world, we must not forget that many parts of nature are finite. We have, over the years, set aside large tracts of land for national parks, wilderness areas, and wildlife refuges—areas that people can visit to get the feel of the great North American continent in its original beauty and diversity. The United States has many beautiful natural areas that should be preserved for all time. We must not be misled into exploiting the lands that have been set apart so that people now and in the future can enjoy them and know what real wilderness is like. We must keep inviolate such natural wonders as the Yellowstone and Everglades national parks, the lovely Allagash in Maine and the mountainous Big Bend deep in Texas on the Rio Grande, and our wonderful system of wildlife refuges in every state. If these areas are exploited, as some people have suggested they should be, all these natural lands will be so depleted and changed that our priceless natural heritage will be lost forever. We must call a halt to the exploitation of these areas. They must remain intact, as they were intended to remain, for the inspiration of present and future generations.

Very early in my career at the Saint Louis Zoo, I accepted an invitation to speak to the Lions Club in South Saint Louis. This was to be a talk about my work with reptiles. I labored over my presentation for a good many days and got together a multipage monograph. On the day I was to deliver the talk, a friend, Dick O'Hare, was in my office. When he learned where I was going he said he'd like to go along and hear the talk. So the two of us drove down to South Saint Louis. Afterward, on the way back to the zoo, I asked Dick what he thought of my talk. He said, "I suppose you want a frank, constructive criticism." And I said, "Yes, indeed I do."

"Well," he said, "first of all, you were speaking way over their heads. You gave an erudite paper on herpetology to some people who knew nothing about reptiles and really didn't care much. That was your first mistake. The second mistake was to read your paper. There is nothing so boring as listening to somebody read a prepared paper as you did. Those people wanted to be entertained, to be amused, and perhaps to learn a little about what you're doing at the Saint Louis Zoo. So in the future, I would suggest you throw away your written paper and speak extemporaneously. You know your subject well enough to be able to give a very good talk."

I followed Dick's advice and have been forever grateful for it. Never again did I read a prepared paper for a talk of this nature.

Other talks followed. I spoke all over Saint Louis to Boy Scout meetings, church organizations, civic organizations, and school assemblies. All this was promotion for the Saint Louis Zoo. As a result of my talks and of the publicity in the newspapers about the reptile collection, one day in the early 1930s I had a call from Jerry Hoxtra of KMOX Radio, who said he would like to come out and interview me. He proposed that a series of live radio programs be broadcast directly from the zoo.

"Well," I said, "I'd be glad to participate in such a thing, but I'd like to get clearance first from Mr. Vierheller." Then I asked, "Who's going to prepare the material for the radio broadcast?"

"Well," said Hoxtra, "I kinda thought maybe you would."

"Why, Mr. Hoxtra," I said, "I don't know anything about radio. I don't know how to prepare for a radio talk."

"It isn't as difficult as you think," he countered. "You just pick a subject, write an outline, and make some notes. You can do most of it just off the cuff."

I asked Vierheller what he thought of this idea and he was all for it. Soon we were broadcasting more or less regular radio presentations directly from the Saint Louis Zoo. We were connected through a telephone line directly to KMOX downtown, but the engineers, the microphones, and the other necessary equipment were right at the zoo.

On one occasion the broadcast originated in the reptile house. I had written a piece about our electric eels. We had three electric eels, each about six feet long, in a large tank on a bunker on the west side of the reptile house. I decided to describe them on radio by asking some people from the audience to join hands and receive a shock when one of their members put his arm into the tank with the eels. I had previously asked Pete French, a Saint Louis businessman who was interested in reptiles and who often went with us on snake hunts in the Ozarks, if he would be willing to be the man who put his arm in the tank. Pete agreed. This action was described and we gave a good word picture of just what was happening. Jerry Hoxtra and George Vierheller were seated at a small table with two microphones, and I was at the very end of the chain of seven people, so that I could describe the shock and ask the other people their reactions. I had written into the script the line, "We'd better be a little careful not to get too close to the microphones, or we might have some electrical interference." I leaned over and put my arm around George Vierheller, who had hold of the microphone stem. I wanted to say that Pete was just about ready to put his arm in the tank and had delivered my line, "We'd better not get too close . . . ," when suddenly the electric eel discharged. We all felt a strong electrical shock in our hands and arms. It coursed right through my body to George Vierheller's body to the microphone and blew the station clear off the air. The big tubes that were used in those days cost four or five hundred dollars. Three of them were blown out at Station KMOX. There was at least a full minute of silence, and that was the last broadcast directly from the Saint Louis Zoo.

It wasn't long before I became involved in Boy Scout work in the Saint Louis area. I acted as counselor for the reptile study merit badge and for zoology. One day George Vierheller suggested we expand this operation. He had been talking to some of the people in the Boy Scout organization and they wanted to have speakers come down to Irondale Reservation, a summer camp for Scouts, in the Ozarks, south of Saint Louis on Highway 21. A meeting was held with Scout officials, and from that we developed a program. I was to drive down to Irondale Reservation on a Thursday afternoon, arriving in time for dinner at about five thirty. After dark I would be introduced at the council fire and give a thirty- to forty-minute talk on reptiles. There was a nature lodge at Irondale Reservation and many of the merit badges in biology were earned by Scouts who studied at the camp. There also was a pit for displaying live snakes and box turtles.

At each of the three sessions during the summer, I wound up my talk by inviting all 300 boys to go snake hunting the following day. After breakfast we assembled and walked down to the edge of a little stream. I gathered all the kids around me and explained what we were going to do and what we expected to see. I told them that no one was to pick up any snake until he had first gotten my okay. There were some copperheads in this area and there is a similarity between the common water snake and the copperhead. I explained this graphically with a diagram on a large piece of cardboard. The dark bands across the body of the copperhead are hourglass-shaped. The dark bands of the common water snake are just the opposite; they are wider along the top of the backbone than they are along the sides, where they taper. And the two species have different-shaped heads. The head of the copperhead is flat on top, with the eyes partially concealed from above by a scale. The top of the head of the common water snake is fairly evenly rounded and the eyes bulge out, much like the eyes of a rat.

We then strung out along the little stream, where we never actually found a copperhead, but did find a number of common water snakes. Pretty soon the boys were catching snakes all by themselves and sometimes they got bitten. But water snakes are not poisonous, and their bite is just a little scratch. If the snakes were big enough, they were taken back to the nature lodge and put on display in the snakepit. At the end of each session the water snakes were released along the stream, and we most likely caught some of them again during the second and third two-week periods.

I didn't then realize the far-ranging effect that those snake hunts would have on the Boy Scouts of the Saint Louis region. Today, I am frequently reminded by Saint Louis businessmen of the times they went snake hunting with me years ago. And so is my wife, Carol. She gives many talks in the Saint Louis area, and lots of people have spoken to her about "the time I went snake hunting with Marlin."

People of the out-of-doors develop a special ability to see and find things. Moody Lentz had this ability for finding snakes. He could see them in places where most people could not. One day we were walking along the edge of a swamp in Arkansas. Through an opening in the foliage we looked out across the swamp, and Moody said, "There's a nice old moccasin sitting on that log out there."

I looked and saw a lot of logs, but no moccasin. "Which log are you looking at?" I asked.

"That one over there," he said, pointing. "Just beyond that light-colored tree."

Looking hard, I could see a speck draped across a log. A moccasin, no doubt! Moody always found more snakes than I did—or anybody else we hunted with. He was just good at it. He had abundant energy, and his eyes were constantly looking and making observations to find snakes.

Another time, while walking along knee-deep at the edge of some overflow water from a river during a flood period, Moody spotted a little snake resting on a log in the shallow water. "There's a nice little moccasin," he said. I took a good look, but the snake was so covered with mud I couldn't see his markings, and I thought his shape looked more like that of a nonpoisonous water snake. I said so.

"Well, I'm not sure," Moody said, "but let's get him by the tail and maybe we can find out." I waded in, grabbed the snake's tail, and quickly lifted him up—and the minute I did so, I saw he was a cotton-mouth moccasin! He tried to crawl back up his body and it looked as though he was going to bite my hand. I shook him down, and handed him over to Moody, who also took him by the tail. Moody swirled him around a few times, and the moccasin started to climb back up to Moody's hand. Moody also shook him down. Then I reached over and got the snake, thinking I shouldn't have handed him to Moody in the first place. The same thing happened again: The snake tried to crawl back up his own body and would have reached my hand, so I had to swirl him around and repeat the whole process until Moody reached over and grabbed him again. Finally, Moody moved near a sloping log, laid the snake's head across the log, pinned him, and caught him. We often kidded each other about passing that hot moccasin back and forth in knee-deep water, not wanting to lose him but not seeing how we could catch him.

One January, a Canadian friend, Al Oeming, invited me to come to Edmonton, Alberta, to speak to the zoological society there. The day after the speech Al invited me to take a drive in the country, promising to show me some snowy owls.

"How many snowy owls can you show me?" I asked.

"Well, how about twelve or fourteen?"

"Twelve will be enough. Let's go!"

We drove out into the country. After a while Al stopped the car, looked off, and said, "Marlin, do you see that little mound way over there in that field?" I said I did.

"There's a male snowy owl sitting right on top of it."

I looked and looked, but I couldn't see any owl on that mound.

Al handed me the binoculars. Then I could see a male snowy owl sitting right where Al had said it was. We drove on down the road, and pretty soon Al pointed out another one. Again I couldn't see it until I looked with the binoculars! There isn't anything really wrong with my vision, which is twenty-twenty, and I was used to seeing things in nature, so finally, after Al had shown me five or six owls, I said, "Al, how in the world do you spot them so quickly?"

"Well," he replied, "I always look for a little spot that's whiter than the snow." That did make it easier, but I never got to be as good as Al at spotting camouflaged birds.

On the morning of the last day of 1928, I was working at my desk in the reptile house when Moody Lentz appeared at the door with our forty-five-inch-long Gaboon viper securely held in his hands. "Marlin," he said, "there are mites on this snake. We better get after them." He walked off toward the kitchen. I followed him, picked up a sponge, turned on the water to make it run lukewarm from the faucet, and mechanically washed the mites off the snake. In those days we had no chemicals that would kill mites without also killing the snake. Today a Vapona ribbon hung in the cage keeps mites under control.

We had just about finished; I was washing the last bit around the side of the viper's face, when he made one of those unexpected lunges, pulling Moody's hand with him, and sank one fang into my left index finger! I immediately put my finger in my mouth and began sucking as hard as I could while Moody dropped the snake into a nearby cage and came back with a scalpel. He quickly made an incision on my finger at the site where the fang had entered. It was on the inside of the finger. Moody immediately began sucking it with his mouth. As he did, we walked toward my office where he could telephone. On the way, I became a little giddy and hit my head on one of the upright posts, which woke me up. When we reached the office, I sat down in the chair and Moody telephoned Dr. Forest Staley, the zoo's medical consultant on poisonous snakebites. Luckily, Staley was at Saint Mary's Hospital, just a short distance from the zoo, and arrived quickly. He put me in his Hudson Super-Six automobile and I was in the hospital within twenty minutes after being bitten.

Dr. Staley took me first into the emergency section of the hospital, made some additional incisions in the top of my hand and applied suction there. We had brought along some anti-snakebite serum from the reptile house, but this was not specific for the Gaboon viper. We were at that time the Mississippi Valley Station of the Antivenin Institute of America. Dr. Alfranio DoAmaral of the famous Butantan Institute in São Paulo, Brazil, had come to the United States and organized this institute for the collection of venoms for the Merck Company in Philadelphia, which later became Merck, Sharp and Dohme. Ironically, I had ordered Gaboon viper antivenom from the Pasteur Institute in Paris, but at the time of this bite it was on the high seas en route to us. I tried to relate my various symptoms to Dr. Staley, and as I did so he would treat me accordingly. George Vierheller put in a long-distance telephone call to Dr. Raymond L. Ditmars, curator of reptiles at the Bronx Zoo in New York, and was in frequent telephone communication with him, getting his advice as to serum treatment.

About an hour after I had been bitten, I suddenly felt that I was going to pass out . . . that it was the end for me. I relayed this to Dr. Staley, who shouted to one of the nurses for some strychnine sulfate, which he injected in one of my veins. This brought me through the critical period when I may have been dying. In a short time I had received two 10-cc injections of the North American anti-snakebite serum, a polyvalent serum against the venom of rattlesnakes, copperheads, and cottonmouth water moccasins. Then, because I began to have difficulty breathing, Dr. Staley realized the respiratory center in my brain was becoming paralyzed from the neurotoxin in the venom of the Gaboon viper, so he injected a tube of cobra antivenin because the cobra's venom is almost entirely neurotoxic in action.

A little later, because of the great swelling and tissue destruction that made the skin of my arm stretch as far as it could clear up into my shoulder and around into my rib cage, Dr. Staley injected a tube of anti-snakebite serum made from the venom of the fer-de-lance. I also had more supportive therapy, including some blood transfusions, and gradually my condition seemed to improve. This happened on the second day after the bite. I stayed in the hospital for three weeks and needed almost six months to recuperate.

Dr. Staley kept records of the case and wrote a case report of Gaboon-viper venomization which was published in the *Journal of the Antivenin Institute of America*. Raymond L. Ditmars reprinted the report in his book, *Snakes of the World*. My recovery was eventually complete, but I still have the scars to show for it, including the scar tissue on my left index finger.

The newspapers and the radio picked up the story. There were daily reports about my recovery. As a result I received many get-well cards, and toward the end of my stay in the hospital, many visitors came by to wish me well. George Deickman, president of the Zoology Board of Control, brought me a whole case of port wine, which was put into the closet of my hospital room. This was pre–World War I vintage, rare in those dry Prohibition days, and he wanted me to take a sip every day to help build up my blood. By the time I felt well enough to take sips of port wine, most of it had already been consumed by my well-wishing friends who had learned about this good supply. Dr. Staley also believed a little alcohol might help my appetite, so he provided me with prescriptions for pint bottles of whisky. This made my apartment a popular meeting place for my friends and in part made up for the fact that I had spoiled their New Year's Eve parties. One of the most touching reactions to my snakebite came from a very good friend, Harig Ruenze, who told me he had made several trips to church to pray for me.

On the day I left the hospital Moody Lentz came to my apartment to tell me that the Gaboon viper had died that same day. He wanted me to know he had preserved the snake carefully in a large glass jar. As soon as I could get back to work, we put the preserved Gaboon viper on display, with a label explaining that this was the snake that had bitten Marlin Perkins. It is still preserved in a storage cabinet at the Saint Louis Zoo reptile house. There also is a picture of the mouth of the same Gaboon viper with his fangs held up. Those fangs are two inches long, the longest fangs of any snake.

Later on, when I was able to resume speaking engagements, I was sometimes introduced with the comment that "it's an interesting observation that the day Marlin left the hospital the Gaboon viper died." This was said with raised eyebrows and a sidewise glance at me, implying that I had something to do with the snake's death. The idea was really shocking to me, and the first time such an introduction was made, I went to some lengths to assure my audience that I wasn't even in good enough physical condition to go to the reptile house that day. I would have much preferred the snake to live.

Many people have difficulty recognizing personalities in snakes. But after you've worked with them a long time you get to know them pretty well. You begin to see personality traits you didn't notice before. For example, in the early days of the reptile house at the Saint Louis Zoo, we had a twelve-foot king cobra, which lived in a cage cared for by Moody Lentz. This snake was quick to raise the forepart of his body, spread his hood, and strike at anyone who had opened the door. To keep the snake from getting out, Moody used to block the door with his body. When the snake paused in an on-guard position, Moody would reach in above him and tap the snake on the top of his head with an opened hand—not enough to hurt, but enough to make the cobra realize that he shouldn't threaten. Moody, of course, fed him as well. Not only did Moody get to know the snake well, the snake got to know Moody. He recognized Moody by sight. Every time Moody passed the front of the cobra's cage and paused to look in, the snake would wake up, hood, or at least follow Moody's movements to the side of the cage. This would happen even on a busy Sunday afternoon when a thousand people were on the floor of the building. As this recognition happened over and over again, it was a facet of snake personality we could often demonstrate to visiting herpetologists and to reporters.

In another cage lived a pair of black cobras from Africa. The larger was about eight and a half feet long and the smaller about four and a half feet long. One morning we arrived at the zoo to find only the larger snake in the cage! The other was missing. We searched the cage carefully. Everything was in order; but there was just one snake in that cage.

We did not close the building to the public, as we were sure the missing cobra could not escape from that block of cages, but we searched the corridors, looked under the cages, and investigated every conceivable snake-hiding place in the building. After a couple of hours of this quest, we thought we had better notify headquarters, so I told George Vierheller that we had only one cobra instead of yesterday's two. We would continue to look, I said, and maybe the snake would turn up. We did indeed continue to look—in every possible spot in that whole building; we wore out the batteries in several flashlights, but found no black cobra. About the middle part of the day, Moody and I decided we had better check one more possibility. We opened the cage door, lifted the remaining cobra on a snake hook, and peered at him critically. I was in front of the cage and Moody was manipulating the snake from behind; I was close enough to see very clearly through the glass cage front. I thought the snake looked suspiciously heavy and full; Moody thought so too. There was a good chance this snake had swallowed its cagemate. This was reported to Vierheller's office. We continued to look for the missing snake, but we were fairly sure our suspicions were right.

Six and a half days later, the large cobra defecated. Sure enough, the droppings contained the scales of his cagemate.

A year or so after that episode, another snake was swallowed by his cagemate. This time both were about the same size. These were chain king snakes from Florida. I happened to pass their cage in time to see the tail of one snake disappearing into the mouth of the other. The remaining snake looked enormous. By that time the reptile house had acquired an X-ray machine through the interest and kind efforts of Dr. John Young, a Saint Louis X-ray specialist, and had devised a cage just the size of the large cassette holding the X-ray film. The chain king snake was placed in this cage and X-rayed. Sure enough, the folded over and looped body of the swallowed snake showed up plainly against the continuous backbone and ribs of the swallower.

Each of these snakes was about five feet long. We X-rayed the cannibal snake every day for about two weeks. At the end of that time, he had completely digested the body of his cagemate. One of these X-ray photographs was placed on public exhibition in the reptile house along with a series of other fascinating X-ray pictures of reptiles, small mammals, birds, frogs, toads, turtles, and fish. Dr. Young had arranged with a supplier of X-ray film to give the zoo its outdated film, which could no longer be sold. We used the gift to produce this large series of X-ray photographs of zoo animals. I can't remember getting anything but good pictures from this outdated X-ray film.

On another occasion we bariumized a rat, tied a string around its neck, and fed it to a timber rattlesnake. Quickly, we took a picture of the rattlesnake which showed the outline of the rat in its stomach and the outline of the gastrointestinal tract of the rat. We took two pictures a day over a ten-day period until the rat no longer showed in the snake. This series gave us pictorial documentation of the digestive processes of the rattlesnake, as the barium was released from the gastrointestinal tract of the rat into the stomach of the snake, then gradually passed down through the snake's small intestine and large intestine, until it was eventually expelled.

Occasionally, a pathological condition would show up in the X-ray photos that had not been noticed at the gross autopsy. This happened with an Indian porcupine that had died and had been autopsied and then X-rayed. The film showed that the porcupine had an impacted gallbladder filled with gallstones. A continuation of the autopsy verified this X-ray finding.

Aside from their diagnostic value, many of the X-ray photos were beautiful in their own right. The picture of a gila monster showed not only the animal's skeletal structure, but also the ossifications in each of the little bumps on its skin. X-ray photos of a bat, a striped bass, the flipper of a sea lion revealed each animal's own fascinating skeletal structure and showed the outline of the soft tissues of those animals. Enough X-ray pictures were developed to allow us to present changing exhibits. That outdated film provided a wonderful opportunity to continue the educational work that is such an important role of any zoo.

7

Jungle Expedition

During the spring of 1927, while the reptile house was under construction, George Vierheller had the idea of sending me to New York for a month or six weeks to study under the direction of the famous Dr. Raymond L. Ditmars, the curator of reptiles at the Bronx Zoo. Vierheller had spoken to Ditmars about this on one of his trips to New York. Everything was agreed upon, and I boarded a train at the Union Station in Saint Louis. Arriving in New York two days later, I spent a six-week period at the Bronx Zoo, mostly in the reptile house, under the friendly and paternal guidance of Dr. Ditmars. I filled a notebook with ideas and suggestions and spent many evenings with head keeper Tumey in the building. Tumey showed me many techniques of handling specimens, persuading them to feed, checking the temperature of the cages to keep them at about eighty degrees, moving snakes from one place to another. I had the privilege of spending long hours with Ditmars and benefiting from his experience and knowledge.

One of the thing Ditmars told me was that snake-hunting expeditions were valuable not only for enriching a collection with specimens you couldn't buy from animal dealers, but also because they generated publicity that could be used to help develop interest in the zoo and the reptiles in the zoo's collection. Ditmars had been to Central America on two or three collecting trips. He suggested this as a good hunting area and gave me the names of people down there who could help.

During the latter part of 1929, when the reptile house in Saint Louis was a going concern, I talked with George Vierheller about the possibility of a foreign expedition. He said the zoo did not have enough money, and suggested I talk to some of the people on the Zoological Society Board to see if they could support such an expedition. I did this, and the money was made available. We started planning what would be the first foreign expedition for the Saint Louis Zoo.

I wrote many letters to Panama, Honduras, and Guatemala. A contact closer to home whose interest turned out to be very valuable was Dr. Willard Bartlett of Saint Louis, a surgeon famous for his goiter operations. Bartlett had made many trips to Central America, from Panama to Mexico, to demonstrate his goiter techniques. He knew many people in Panama, Honduras, and Guatemala. In the course of our discussions about the trip, Bartlett decided to go with us. This was wonderful news, and it gave us a firmer base from which we could organize many of the details of our forthcoming travels. As it turned out, Bartlett was a wonderful traveling companion.

In order to ship the specimens back, I purchased two large, black, trunklike fiber cases and had the carpenters at the zoo arrange removable trays of pressed wood inside the cases. When I went down to inspect the cases, the carpenters working on the job suggested that the bottom position in the tier of compartments would be an excellent hideaway in which to bring back some good booze. Of course this was during Prohibition and good whisky or gin was a luxury. Strangely enough, I had thought of the idea myself, and it shocked me a little to find that somebody else had come up with the same concept. Later, when Vierheller came by to check on the construction, he had the same idea! We showed those cases to some of the members of the Zoological Society and were surprised to find that they too thought this a splendid way to bring in some liquor. Bartlett agreed, and so did George Pringle and his wife, who were aboard the United Fruit Lines ship *Castilla*, on which we sailed from New Orleans for Panama. And so did the captain of the *Castilla* and several of his crew members. If everyone had the same idea, it really might be possible.

The *Castilla* was the first ocean vessel I had ever been on. Moody and I shared a small cabin. Because of Dr. Willard Bartlett's status, we were invited to join him at the captain's table. Each evening before dinner we were welcomed in the skipper's quarters for drinks. There I tasted my first martini, which had a pickled onion in it. I thought this was the worst thing I'd ever tried to drink in my life, and I couldn't get it all down. On subsequent evenings during the four-day trip we were served the same drink, and I must confess that toward the end martinis tasted a little better than they had at the beginning, but not much.

We landed at Ancon, on the gulf side of Panama, and took the railroad that parallels the Panama Canal over to the Pacific side. I was surprised to learn that the canal runs north and south instead of east and west. We arrived at Panama City and checked in at the Tivoli Hotel.

In Panama City we met an old friend of Bartlett's, Dr. Clark of the Gorgas Memorial Laboratory, a man who had spent much of his life investigating the health problems of the tropics, and particularly of the Canal Zone. Through him we learned of a highway through the jungle to the Madin Dam, which was under construction. All arrangements had been made, and we took the train back along the canal to a place called Frijoles. There we were met by one of the Americans in the construction crew and some Panamanians. They had the dugout canoes that were to be our transportation. We loaded our equipment, which included two recently purchased folding canvas cots and a couple of cotton blankets. The boats had outboard motors, and we were soon working our way upriver and then finally up a smaller stream to the site where the roadgang was building the highway. We ate in the mess hall of the construction crew, so didn't have to cook or take any food with us. We were free to hunt both night and day for specimens.

It felt good to be in the tropics, in a real rain forest, and to experience the unusual odors and the sounds of insects and frogs new to our ears. Moody and I felt as though we were in a true paradise. After dinner the first night, we set up our cots and turned in. In a short time we began to feel cold through the canvas of the cots; by flashlight we looked around and found some papers, which we put underneath the blankets. A little warmer, we managed to sleep. We had never expected it to get so cold at night in the tropics.

We were up quite early, then hunted up the creek where we felt we couldn't get lost. Moody Lentz and I had had much experience collecting reptiles in Missouri, Illinois, Arkansas, and other parts of the United States. "Boy," we thought, "here in the tropics we'll really be able to collect not only tropical reptiles, but many of them."

We used our customary technique of turning over logs and rocks. We spent the whole day looking for specimens and saw not a single snake we could collect for the zoo back in Saint Louis. What an inauspicious beginning for our first foreign collecting trip!

The second day we caught a few snakes, though nothing spectacular: a sumebadora, which is related to the indigo snakes of the southern part of the United States, and a couple of vine snakes, long slender creatures that resemble thin pieces of vine and feed on small lizards. But that was not what we would call a very productive collecting day.

After dinner that second night Moody and I decided to see what night collecting was like. As soon as darkness fell, we got our flashlights, snake sticks, and snake sacks (which we hung around our belts), and moved off a ways from the campsite. We were going up a little hill, peering as best we could through the tangle of growth, when suddenly we heard something. We weren't more than six feet apart and we both switched off our flashlights at the same instant. We waited, not moving, just listening. Whatever it was, was coming through the dry leaves making a big racket. Through our minds flashed the thought that the most dangerous thing we could encounter was a jaguar. Yet we didn't think a jaguar would be dragging its feet that much. Other possibilities included a tapir, a capybara, and a number of other mammals. The noise got closer and closer and closer, and reached a point where we couldn't stand it anymore. Without saying a word to each other, we both flicked our flashlights on at the same moment, and there was an armadillo. He was headed right for me, and as he passed over my foot, I simply reached down and picked him up by his tail. Another fine specimen and the first mammal collected on the expedition.

We spent three days hunting around that road camp, but found very little. We were two disillusioned snake hunters when we left. Jungle hunting was nothing like snake hunting back home.

On the way back we stopped off just opposite Frijoles at the famous Barro Colorado Island, at the invitation of Dr. James Zetek.

Dr. Clark had introduced us to Dr. Zetek in Panama City. Barro
Colorado Island was then maintained in part by Dr. Thomas Barbour
and Harvard University, but later came under the control of the
Smithsonian Institution. Originally a high hill, Barro Colorado became an island when the water rose to form Gatún Lake, which is
part of the Panama Canal. As it rose the water forced many animals to
take refuge on the hill. All the animals in the surrounding valley that
could reach this high spot were then trapped on this interesting island. A laboratory had been built there, and quarters for visiting students. While we were at the island Dr. Robert Cushman Murphy of
the American Museum of Natural History was there too. We were
given permission to take a few reptiles on Barro Colorado if we could
find them, although normally hunting was not allowed. A local assistant named Pedro was assigned to us, although there were well-
marked trails throughout the island and you couldn't get lost for very
long.

It was on Barro Colorado Island that Dr. Ray Carpenter made his
famous observation of the behavior of black howler monkeys. He
found that one of the reasons for the loud calling of these animals is to
maintain contact with neighboring troops of black howlers, so that
troops can meet soon after dawn and hoot and howl at each other and
thus, by threatening each other, rid themselves of aggression within
the group.

Soon after arriving on the island, we took to the trails and before
long came to a downhill section of Agassiz Trail, named after the
famous Professor Agassiz of Harvard. (All the trails on the island were
named after scientists.) As we jumped across a small brook at the
bottom of the trail, we saw on the hillside above us a great troop of
white-lipped peccaries, much larger animals than their cousins, the
collared peccaries. A big boar came forward clacking his teeth together and making threatening gestures. For our part, we had frozen
in our tracks. While the boar was threatening, the troop turned to our
right and filtered off through the undergrowth. The boar made one
dash forward, to cover about half the distance between us, then
abruptly turned and followed the rest of the troop. I had a camera in
my hand at the time, but was so intent on watching the actions of the
boar that I didn't think of taking a picture.

I looked to see where my companions were. Moody was standing
slightly to my left and back of me. On the other side was Pedro with

his hands on a devil thorn—a tree that has sharp-pointed conical spikes all over the bark. I assume he was prepared to climb that tree if the peccaries had made a real charge.

Later on, while following another stream, we heard a slight rustling in the leaves. Moody was in the lead on this occasion; he signaled to me that he was watching something, and as he watched he slowly backed up, never looking away from that spot for very long, moving back to where I was. Then he whispered, "There's a jaguar sitting on the top of that bank!" We were unarmed, and we didn't linger.

We moved off to another trail and continued our hunt. We saw many birds and were delighted with the thrill of watching Oropendula fly to their long, hanging-grass nests and enter from the bottom. When we returned to camp, I immediately told Murphy about our encounter with the white-lipped peccaries. He said, "Boy, are you sure you saw a white-lipped peccary? They have never been reported before on this island." I said, "Yes, sir. I'm quite sure about it. We have collared peccaries in our zoo; I've seen pictures of white-lipped in the literature. These were much larger animals and they did have white hair on their upper lips." He gave me a disbelieving look, but in a report from Barro Colorado Island the following year, Dr. James Zetek also reported them. I had mentioned our sighting to him when we got back to Panama City.

When we returned to that city Bartlett, who had been demonstrating his famous goiter operation at one of the hospitals, had good news for us. He had met the president of Panama and the president was going to send us down the Tuira River almost as far south as Colombia on his private yacht. The yacht was going down anyway on a patrol mission and we were more than welcome to go along. Early the following morning, having purchased our own food and a few utensils for camp cooking, Moody and I arrived at the pier where the yacht was tied up. It certainly did not look like my idea of a private yacht. It looked more like a small tugboat or maybe even a small harbor-inspection boat. There was a crew of two, and four members of the military were also on board. We were assigned two bunks directly behind the diesel engine and over the shaft. The others were to sleep on the forward deck on their blankets. With everything stowed aboard, we set off, practicing our high-school Spanish on the crew members and soldiers. We moved out past the fortified hills of the Pacific entrance to the Panama Canal, and with the Pacific shoreline of Panama in sight most of the time, we cruised southward. There were some islands on the starboard side and I looked for the yellow-bellied sea snakes that are known to occur in this region. For meals, we opened a couple of cans, and that evening after dark we retired to our bunks. We didn't stay there long; the fumes and vibration from the engines and the propeller shaft drove us back on deck. Those annoyances would have made sleep impossible and maybe even dangerous. We finally found deck space up forward where we could sleep.

Late in the afternoon of the second day, we turned into the broad mouth of the Tuira River, which was about two miles wide, and proceeded upriver for some hours before reaching our destination—a little village called La Palma. There the boat docked. Moody and I, with our luggage, food, snake sticks, snake sacks, and flashlights, hurried ashore. The boat didn't linger, as the passengers had work to do upriver, but they promised to return in three days' time. They also helped us find a place to stay above a little grocery store. We climbed the steps on the outside of the building; our quarters was just one large room with no furniture. We had our canvas cots and our cotton blankets and a couple of boxes of food, so we thought the accommodation would be fine. We walked up and down the one street of the town, which paralleled the bank of the river, and noticed that all the houses were on stilts because the river rose and fell with the tide. I suppose that made it convenient to throw garbage out the back door and have it disappear into the river. A number of black vultures sitting on the corrugated tin roofs of the houses helped dispose of the garbage.

The one street was only about a block long and there were a couple of thatched huts at one end. All the small children were naked and appeared to us to be a mixture of different bloods. One boy had strong Chinese features; there was Indian and Spanish and East Indian blood, and we could see a scattering of Negro. The mayor of the village came by, introduced himself, and invited us to join the following evening in a small fiesta, as there was to be a dance. We accepted with pleasure.

We engaged a man to take us across the river, where we had been told there were many more snakes. But the hunt would have to wait for tomorrow, as the sky was now getting dark. We went back to our room to see what kind of meal we could put together. There was no place to light a fire, so we opened a few cans and ate the food cold, or at least started to, but it needed seasoning and we hadn't bought any salt.

"Don't worry," I said to Moody, "I'll go downstairs and get some in that little grocery store."

Down the stairs I went, and there was a woman. "Buenas tardes, señora," I said in my best Spanish. "Yo quiero algunas sal."

"Qué?" she said.

I said, "Sal."

Again she said, "Qué?" So I made some salt-shaker motions with my hand, repeating the word "sal."

She didn't understand. I looked around and there on a shelf was some salt; I reached up and got it and said, "Este."

She said, "Ah, sal!"

I really couldn't tell the difference.

The best part of our meal that evening was the canned peaches.

We paddled across the Tuira River the next morning with our guide and started hunting snakes on the farther bank. It was hot! And humid! And no snakes to be seen. We were about exhausted when our guide stopped and asked us if we were thirsty. We were. With his machete he cut long sections out of a vine, sliced them open, and handed them to us. Water trickled from the vine into our open mouths. But we had no luck with snakes in this locality.

That evening we again walked down the village street and noticed a small bar. We went in, sat down, and ordered two drinks. We were talking about being the only gringos in La Palma when the door opened and in walked three American sailors. They too sat down at the bar and ordered drinks. I said, "Say, you fellows are Americans, aren't you?" And one of them said, "Sure."

"Where did you come from?" I asked.

"Oh," he said, "we are attached to a station out there on an island."

"What kind of a station is it?" I asked.

"It's a naval station," he said. "We're guarding the entrance to the Panama Canal. It's mostly radio."

So we weren't as isolated in an out-of-the-way village as we had thought. Later that evening we went to the dance and had a pretty dull time, until Moody, who was dancing with the mayor's daughter, decided to teach her the Charleston. She learned the spectacular dance quickly and did it well. They were a sensation.

The next day the president's yacht came to retrieve us, and we had a rough passage back to Panama. The decks were awash virtually all night; we were soaked, as was a lot of our gear.

From Panama we took a boat to Puerto Barrios in Guatemala. This is the principal port on the gulf side. It consisted of four streets, each about two blocks long, lined with one-story houses. Behind the houses the village gave way to banana plantations. Puerto Barrios was also the terminal of the narrow-gauge railroad running westward through the banana plantations and the jungle, climbing as it continued on its way to Guatemala City.

We cleared customs and boarded the train for our next base of operations at Quiriguá. Here the United Fruit Company maintained a fine hospital headed by an old friend of Dr. Bartlett, Dr. McPhail. Surrounding the hospital was a beautiful, well-kept, nine-hole golf course. We were given nice rooms and a place to keep the specimens we hoped to collect there. Nearby was one of the oldest Mayan ruins, also called Quiriguá. We were naturally anxious to see the ruins, all the more so when the local people told us they frequently saw snakes there. So our purpose in visiting the ruins was twofold. Transportation locally was by the narrow-gauge railroad, and Dr. McPhail put at our disposal a putt-putt unit that ran at our discretion, depending upon the position of the trains, which were not numerous. At the entrance of the historic area was a large sign: These Ruins Have Been Excavated by the Saint Louis Archeological Society. There were some very attractive stellae. The temple was not completely excavated, but you could discern the outline and see some of the walls and the stairs leading to the lower level. Cattle grazed through this area, and that helped keep down the foliage. Near the ruins we caught three interesting snakes—very dark green with emerald green spots all over them. They were fast, and we had to be quick to run them down. They were almost as fast as our coachwhip snakes.

Another trip on the narrow-gauge railroad took us to the Zacapa Desert. Though we got to the edge of this arid expanse fairly early in the morning, and hunted all day, it was so hot that the snakes were not out.

From Quiriguá we went back down the railroad to Puerto Barrios and caught a ride on a United Fruit boat to Tela, Honduras. At Tela there was a serpentarium right next door to an experimental tropical-fruit farm. Dr. Pomponoe was in charge of the farm, and Ira George ran the serpentarium. We collected nearby and found a large freshwater snapping turtle, similar to our American snapping turtle, but bigger and darker in color. We also caught some fer-de-lance snakes and some boa constrictors.

From Tela we went to San Pedro Sula, where Dr. Waller, one of the men Raymond Ditmars had told me to contact, had reservations for us at the one and only hotel. Waller was an American medical doctor from Louisiana. He had lived in San Pedro for a long time, married a Honduran woman, and had quite a number of children. He was interested in reptiles, made one of his employees our guide, who showed us areas where he had frequently seen snakes. This proved to be one of the best snake-hunting areas we had so far encountered. Green iguanas were abundant. We collected quite a few by climbing trees, shaking the animals out, and, when they jumped to the ground, grabbing them before they could recover their wits and run for safety. Some iguanas jumped into a stream and then we had simply to wait until they surfaced, wade out, and make the catch.

Ditmars had told me about a marvelous place for collecting crocodiles, Lake Ticamaya. When we asked Dr. Waller about it, he said it was well known and introduced us to a banana planter who could help us get there. This man had been educated in the United States and spoke English. He picked us up at three o'clock one morning at the hotel and drove us in his truck to Lake Ticamaya. As soon as we arrived, we met some of the children of a man who had a reputation as a successful crocodile hunter and who lived on the bank of this lake. Our friend introduced the children and asked one boy, a ten-year-old, to take off his shirt, telling us this youngster had quite a history. When the boy was about five, he and his seven-year-old brother were playing in the water in front of their house. A crocodile grabbed the younger child by his chest and started to swim off with him. The seven-year-old picked up a stick, ran into the water, and beat the crocodile over the head until it let go of his brother. The big crocodile had injured the boy's chest severely. His chest was flattened from the pressure and you could see distinctly where every tooth had entered the child's skin, on both the front and back of his thorax. The boy was, however, completely recovered.

The children's father was going to take us crocodile hunting that night. At dusk he and two of his older boys showed up in a couple of dugout canoes. Dr. Waller and our banana-rancher friend had told us that this man was a Dead-Eye-Dick with a harpoon. He could hit a crocodile anywhere we wanted. We asked him to hit the tail so it wouldn't injure the animal badly. We could then pull in the crocodile, hog-tie it, and apply a healing ointment to the small puncture made by the harpoon. Moody and I sat in the boat with the harpooner; the older son paddled. Another son was in a backup boat behind us. We had headlights and soon picked up the red eyes of crocodiles lying submerged with just their heads sticking up above water level. We eased up to one of these animals, and from about ten feet away the hunter hurled his harpoon. He missed by a good four feet. Using the line attached to its end, he pulled the harpoon back and hooked it once more onto the throwing stick. We approached another crocodile. Another miss. The third or fourth throw made contact, but with the animal's shoulder, not his tail. The crocodile immediately started thrashing around in the water, turning over and over; we could hear those mighty jaws snapping together. We eventually tired him out, brought him alongside, put a noose over his jaws, grabbed him, brought him aboard, and hog-tied him with his legs up over his back. His tail was bent around a little so he couldn't get enough leverage to slap with it. He was then put in the other boat and taken to the bank, where our truck waited. We finally caught three crocodiles—one about five, one about six, and one about seven feet long.

When I asked our Honduran friend why the harpooner had been so badly off target that night, he said, "Oh, I forgot to tell you. He went to a wedding about two weeks ago and still hasn't recovered."

We put the crocodiles in the banana rancher's truck and headed for town. We arrived around midnight. The hotel was dark and quiet; Moody and I slipped up the backstairs to our room. The hotel manager had asked us not to take any animals into our room, but to place them in his spacious backyard. He had in that yard a number of Manx cats, and we were afraid that if we just laid the crocodiles on the ground the cats might gnaw at them. We decided to house the crocodiles in our room. We carried them up very quietly and placed them on the floor. It was a spacious room with good ventilation, but very thin walls between it and the rooms on each side. One bed was next to one wall, and the other next to a second wall; there was a large open space between. We were dead tired, as we had been up since three o'clock in the morning and it was now twelve o'clock the next night. But the crocodiles were pretty thoroughly tied, and we didn't anticipate any problems. We were just dropping off to sleep when we heard a *thump, thump, thump.* A flashlight showed a crocodile testing his ropes and trying to move. As soon as the light came on, he quieted down. We tried to go to sleep again, and again, just as we were drowsing off, *thump, thump, thump* went the crocodile until we flashed the light. Once more, the crocodile quieted down, and again we tried to sleep. *Thump, thump, thump!* It went on all night.

We decided we'd better get those crocodiles out of the room before the hotel was awake for the day, so at six o'clock we started to carry them downstairs. One had moved a little ways under my bed, so I reached for his tail, and as I pulled him out saw that his jaws were no longer tied. Somehow he had managed to squirm his way out of the rope holding his jaws and head. More activity, as we managed to tie him up again. We carried the crocs on our shoulders down the stairs and into the backyard and penned them for the time being in a nice fenced-in area. Through the owner of the hotel, we located a carpenter shop not far away, and ordered a crate to be made for the crocodiles and for some of the turtles we had collected. I asked the carpenter what kind of wood he could use, and he said, "The cheapest wood we have is mahogany." So we acquired a crate fashioned of solid mahogany boards, about eight and a half feet long, four feet wide, and eighteen inches high, with a special compartment at one end for the turtles.

The day before we were scheduled to depart we began installing our collection from Panama, Guatemala, and Honduras in our new mahogany crate and the two big black fiber cases I had had constructed at the zoo back in Saint Louis. Looking over those cases with their removable trays, I remembered how everyone had been telling us what a great place the bottom compartments would make in which to conceal some contraband liquor. I felt a bit nervous, however, and finally decided that if all those would-be smugglers had had the same idea, the customs inspectors would likewise realize the value of a case of snakes as a hiding place.

As a result we came back absolutely clean, with no liquor of any kind anywhere. We loaded our collection, housed in the three containers, onto trucks, drove to the harbor of Tela, Honduras, to put them aboard our United Fruit liner, and shortly afterward sailed for New Orleans.

When our ship landed in New Orleans, we were met by Bob Safford of the *Saint Louis Post-Dispatch*, who had explained to the customs inspectors the desirability of getting our cases on the first fast train back to Saint Louis. So when the time came, the inspector asked, "What do you have in those two big black cases?"

"Snakes and lizards and a couple of small turtles," we said.

He quickly stamped the cases and the crate and said, "Okay. Get them out of here. We sure don't want to go rummaging through those!" A great opportunity had been lost!

We arrived in Saint Louis to much publicity in the newspapers and on the radio. For about a month a special exhibit on our expedition occupied one wall of the reptile house.

I had hoped to salvage the mahogany crate and use the wood to build some furniture. But the first days after our arrival were so occupied in developing the exhibit that I didn't get around to my crate for some time. Finally I asked about it. "That old dirty box the crocodiles came in?" our maintenance foreman asked. "We had it out in back of the maintenance barn for a while, but in cleaning up a couple days ago, we burnt it!"

8

The Performing Primates of Saint Louis

It has long been the practice at the Saint Louis Zoo to take young anthropoid apes outside in good weather to play on the lawns in the sunshine. One day, primate trainer Cash Ferguson had Yannah, the young gorilla, out for exercise on the lawn between the reptile house and the primate house. She started up a forty-foot tree. This seemed fine exercise, so Cash let her go, and soon we noticed that she was beginning to make a nest. She selected a limb on which to sit, pulled branches toward her, and tucked them in under her feet. Soon just her shoulders and head were sticking out of the nest she had constructed. I ran and got my camera to record the performance.

We wondered what the other youngsters would do. Cash then brought out a young orangutan. This animal also made a nest, using very much the same technique—pulling the branches toward him and tucking them in underneath. This ape, though, broke off some branches and placed them over his shoulders and head. A young chimpanzee was given the same opportunity in another tree. Seeing the gorilla and the orangutan building their nests, he did the same, except that he broke off most of his branches and tried to make a platform on the branch he was resting on.

It is interesting that these apes, separated genetically and geographically, are able to build nests when they have had no previous experience in such construction. Perhaps each had some memory of riding on its mother's back when she built a nest; or of seeing other troop members build nests in trees in the evening. But whatever they remembered had happened pretty far back in their lives, as all these animals had been captured as infants.

Robert Yerkes, in his book *The Chimpanzees*, describes experiments to determine chimps' ability to reason out a problem. He placed the animals in a large room with some food hung above their reach on a chain, and various boxes scattered around. The chimps quickly discovered that if they piled up the boxes they could climb up and reach the food which they couldn't reach any other way. This experiment sounded interesting, and we decided to repeat it at the Saint Louis Zoo. There was a large cage for the exhibition of these animals in the regular chimpanzee show, and the stage had a nice high ceiling. From a ring in the ceiling we hung a short chain and to this we tied some bananas. Six or eight wooden boxes were scattered around the floor. Cash took two chimps, Sammy and Billy, into the cage, showed them the bananas, and left them there. We watched through the bars from the front and side doors. The chimps played and stomped around the cage for a while; they climbed up the front bars. Suddenly Sammy stopped, looked straight at the bananas, and obviously got an idea. He pulled a box directly beneath the bananas, climbed up on the box, and reached upward. He was still a long way from the bananas. He got another box, put it on top of the first one, climbed up, and reached. Still too far away. He piled three, and then four, boxes; the fourth fell off and he had to get it back in position. He whimpered and cried, but he stayed with the project (getting no help from Billy, who was just running around the cage and playing). Finally, although the boxes weren't stacked carefully, Sammy was able to climb on top and reach a banana. He tore it loose from the rope and jumped to the floor.

This experiment was repeated many times, and it was always Sammy who got the banana. One day Cash laid a tall pole at the back of the cage before he took the animals in. As usual Sammy got a box and placed it directly under the bananas. He got a second box and placed that on top of the first one. And then he saw the pole. He ran over and brought it back, pushed the boxes away, set the pole down directly underneath the bananas, quickly ran up the pole, grabbed the chain with one hand, hung on to the pole with his hind feet, tore a banana loose, put it in his mouth, and then, steadying the pole slightly, slid down the pole. This was a quicker way than piling up boxes.

Before the primate house was built at the Saint Louis Zoo, the lion house contained not only the big cats, but a few other animals as well. There was, for example, an orangutan known as Sam. He got his nickname, "Spitting Sam," because of his ability to spit on visitors from a great distance and with great accuracy. Max Mahl was in charge of this building, and because he had had some experience with chimpanzees and orangutans while working in a circus in Germany, he had a close personal friendship with Sam. Max used to take Sam outdoors for exercise and taught him to ride a bicycle, to carry a little gun, and to salute. Sam was the first of the performing animals at the Saint Louis Zoo.

Sam was followed by three young chimpanzees, Mike, Henry, and Duffy, also trained by Max Mahl. When the primate house was finished, there had been built into it a large cage that became the performing arena for the chimpanzees and the orangutan. They did a variety of acrobatic stunts and would conclude with a teaparty, drinking their tea from cups just like people. When I knew him, Max was an older man, with white hair and long, flowing, white mustaches. He had a standard reply when people complimented him on his chimpanzee show and asked, "How in the world, Max, were you able to train those chimpanzees?" He would respond by saying, "Vell, you got to be just a little bit schmarter." When Mike, Henry, and Duffy grew up, as chimpanzees inevitably do, and had to be retired from show business, Max Mahl retired also.

Max was followed by Cash Ferguson as trainer. In Cash's show Sammy and Billy first rode tricycles around the stage, then graduated to bicycles. They also did a wonderful whirlaround on roller skates. The show always ended with a boxing match. Cash dressed the chimps in boxing trunks and boxing gloves and placed them in a small ring with a genuine bell. They started round one. There was the usual sparring, getting in position, swinging a haymaker, missing, and the bell was quick to end round one. In round two Sammy seemed to come to life and started whaling the daylights out of Billy, who, upon

a cue from Cash, fell over on his face. Cash counted Billy out, held up
Sammy's hand as the winner, and Billy automatically got up and left
the ring. He knew the act was over.

The first gorilla ever exhibited at the Saint Louis Zoo was the
young female Yannah, who had come to Saint Louis from Carl Hagenbeck's animal emporium in Hamburg, Germany. Arrangements were
made to have her shipped to the United States on the *Graf Zepplin*.
George Vierheller met her at Lakehurst, New Jersey, and brought her
to Saint Louis from New York on an overnight train. As he always did
when traveling with young chimps and orangs, he took a compartment in a Pullman car. Yannah and George had their dinner and rode
throughout the night in the privacy of their special room. We often
wondered what the other passengers would have thought if they had
known their next-door neighbor was a gorilla.

There was much publicity about Yannah's voyage, and Saint
Louis was delighted to be one of the few cities in the world to exhibit
this rare anthropoid ape. She was gentle and friendly, and it wasn't
long before Cash Ferguson had worked her into his act. One part of
the act was a version of German gymnastics for exercising and performing acrobatics. Sammy and Billy were pretty good at gymnastics,
but Yannah didn't respond as well, so wound up wearing a pointed
dunce cap. Cash Ferguson and George Vierheller used to discuss the
relative intelligence of these two different kinds of anthropoid apes. I
remember Cash Ferguson remarking, "Well, I don't know whether
she's really dumb or just plays dumb so she doesn't have to work."

The first orangutan to be born at the Saint Louis Zoo was a pretty
little female named Patty Sue. Her mother didn't take care of her and
she was soon put on a bottle. Because she had to have regular around-
the-clock care, Cash Ferguson used to bundle her up in a baby's blanket at night and take her home with him where he could bottle feed
her and keep her warm, between himself and his wife when they went
to bed.

Leon Smith succeeded Cash Ferguson as trainer of chimpanzees.
It wasn't long before a special cage was constructed near the south
waterfowl pond, near refreshment stand number two. Visitors could
stand around this cage and watch the performance. There was tightrope walking, walking on stilts, and even an orchestra. But best of all
was a magic act. An assistant passed a slate around the crowd and
asked a child to write a number between one and ten. The slate was

put in a bag with a drawstring, and passed by the assistant to Leon Smith, who placed it in a box and closed the padlock. The chimpanzee Jackie was the mind reader who would come up with that number. Of course, he couldn't talk, but there was a way around that. Jackie stepped up to a board that had on it a bell with a pull cord. He would concentrate by putting his hands over his eyes. When Leon said, "Have you got the number? Are you ready?" Jackie would nod his head. Leon would stand at attention, motionless as a soldier, looking past the animal at the audience. Jackie would ring the bell six times (or whatever number was on the slate), back away from the bell, take his seat, and wait for the results. Leon would get out a key, open the padlock, reach down into the box, bring out the cloth bag holding the slate, take out the slate, and hold it up so that all could see the number was six. Jackie had come up with the right answer. Of course, everyone was mystified and wondered how the trick was done. So far as I know, no member of the audience ever figured it out. If you promise not to tell, I'll let you in on the secret. The assistant who retrieved the slate from the child who had written the number, of course, knew the number. When he passed the slate inside the bag to Leon Smith, the correct number of fingers were on top. Leon then knew the number. When Leon took his military stance, looking past the chimpanzee and the audience, Jackie could see his eyes. Leon was staring straight ahead. And when the right number came, all he had to do was move his eyes, and Jackie stopped ringing the bell.

When the anthropoid ape house was constructed at the Saint Louis Zoo in 1940, an arena for the performing chimpanzees was built lower down the hillside. Fitted into the naturally curving slope were bench seats for 500 people. Part of the arena was covered over by a roof, and all seats faced the stage, which had no bars—just two upright posts to support the roof. The stage was surrounded by a water moat, and along the edge of the wall of the moat nearest to the stage was installed electric wire such as is used in electric fences for cattle. This wire extended up along the edge of the roof, so that if a chimpanzee tried to climb out from the top of the two poles supporting the roof he would get a shock. One shock was all an animal required; he never tried again.

The stage was virtually round, with two doors in the back through which the animals were brought on. On the audience side of the moat was a living hedge to soften the effect of the harsh concrete walls. The performing chimpanzees, which by this time numbered about twelve, lived in the exhibition cages in the anthropoid ape house. This was where they had their meals and where they slept. Leading from the inside of the anthropoid ape house was a tunnel that passed below the seats, underneath the stage, and up a ramp to a large room just in back of the stage itself.

The trainer at this time was Mike Kostial, an active, good-looking man who resembled Clark Gable and like that fine actor was himself a great showman. He came by this naturally; his father had been an animal trainer, in Germany and in the United States, first at circuses and eventually at the Saint Louis Zoo, where he trained the lions, tigers, leopards, and bears. During the time his father was the trainer for the lion and tiger show, little Mike was frequently at the zoo and became well known to all the people who worked there. When he finished his education, Mike was ready to step into his father's shoes and start training animals.

Each year a theme was chosen for the chimpanzee shows and this was depicted in the rear mural, which was painted by Mike Kostial himself. The props were likewise designed by Mike and built by him and the zoo's construction department. One year the theme would be the Wild, Wild West. Another year, Circus Days. The costumes had to be designed to fit the theme. Each animal had to be measured and fitted with his costume. This was done by Mrs. Floyd Smith, the wife of the trainer for the elephant show. Mrs. Smith also made the costumes for the elephants.

George Vierheller started almost from scratch and built up one of the most important zoos in the world. He was a wholly dedicated man who loved his zoo and wanted everybody else to love it also. It all started in the early 1900s, when a few zoo-minded, public-spirited citizens, such as George Deickman, Andy Bauer, and Frank Schwartz secured from the state legislature in Jefferson City an enabling act which made it possible for the citizens of the city of Saint Louis to vote a two-mill tax to support the zoo. This two-mill tax (two-tenths of a cent on city taxes) was passed in 1914, and soon afterward, George Vierheller became secretary to the board. Previously, he had been a telegraph operator in a brokerage office, but it soon became apparent that the zoo needed somebody to manage it properly, so George Vierheller was given a second title, superintendent, as well.

After World War I, enough money had been amassed to build a major exhibit. John Wallace was engaged as architect, and he laid out the beautiful vista called Peacock Valley, with its series of ponds ending at the sea lion basin. Wallace and a man in Denver who had developed a new system of artificial rock construction laid out the design for the bear pits. The method of making artificial rocks involved taking plaster casts of rock faces on the limestone bluffs of the Mississippi River south of Saint Louis. The four- by six-foot plaster casts, like photographic negatives, could be copied and then cut to any shape a wall required.

The new bear pits had a moat twelve feet deep made of smooth concrete that the bears could not climb, and a small wall of the natural-appearing rock to partially hide the moat from viewers, who were kept back by an ornamental guardrail. Living plants made a pleasing effect between the guardrail and the low wall. Vines were planted to overhang the top wall of the pits, and behind that rose a forest of trees and shrubs. There were water pools in the pits where the bears lived and spacious dens behind where they could spend the nights in the cold weather, and have their cubs.

Work on these pits was started in 1919 and finished in 1922. The design paved the way for other barless exhibits, not only in the Saint Louis Zoo, but in a number of other zoos in the years to come.

At this stage, not requiring all of the income for operating expenses, George Vierheller would allow $200,000 or $300,000 of tax money to accumulate, and then he and John Wallace would develop new designs for other animal exhibits. The primate house followed the bear pits in 1924. Its exterior was tan stucco with a red tile roof in a lovely Mediterranean style. The entrance columns and the windows were trimmed in terra cotta with animal motifs. Upon entering the doors your first view was of a very large cage bathed in daylight from the overhead skylight and set off by a planted area of shrubs and small trees surrounding it. Bordering the planting were white columns with capitals of carved plaster colored with burnt sienna to better emphasize the design. Between the white columns around the planting was a beautiful wrought-iron railing, ornately designed. The design of this beautiful centerpiece, a patio setting with living monkeys as the chief point of interest, became a distinctive feature of the Saint Louis Zoo. The concept was carried over to the reptile house built in 1927, and then to the bird house constructed in 1930. Both buildings are of

Mediterranean architecture. The terra-cotta trim and ornamental rail-
ings featured designs of reptiles and amphibians for one and birds for
the other. The concepts of the overall design of these buildings were
copied in modified forms and used by other zoos.

The bird house was the first of its kind in the world in which the
patio was designed as one large cage with tropical planting and a
stream flowing into a pool. The planting area and pool were depressed
about four feet below the floor level. As visitors entered the inner door
from a large vestibule, the view of this big flight cage was uninter-
rupted, for there was no glass or wire. The guardrail kept the people
out, but nothing kept the birds from flying around the public spaces.
(They did fly off on occasions, and for several years superb starlings
from Africa nested inside the amber glass of an ornamental chan-
delier.) Standing at the railing one could see a variety of tropical birds
in the patio cage, then look through that to another cage, also planted
with tropical foliage, and on through the glass, across the public space
and into a large glass-front cage with a natural-habitat mural painted
on the canvas of the large curved back wall. Looking obliquely, left
and right, one saw other cages, some with glass fronts.

The fabulous antelope house was covered with replicas of the red
granite boulders near Graniteville, south of Saint Louis, at a place
called locally Elephant Rocks. From a visitors' walkway one viewed
antelopes, giraffes, or zebras in spacious yards, separated from the
public by moats. When looking across a moat, one could not see that
there was a complete building under that great pile of granite boulders. Even the chimney was hidden by a boulder completely surrounding it. The separation walls of the outside yards were also of red
granite. Three satellite buildings nearby were also covered with artificial red granite boulders, and yards radiating from these matched the
moated yards of the antelope house. Visitors had a clear view of the
grevy's zebras, Bactrian camels, black rhinoceros, aoudad sheep, black
buck antelopes, gaur buffaloes, wildebeests, and red kangaroos in spacious areas with no bars or wire to interfere.

The antelope house at Saint Louis can be entered from either
end, and the moment you pass through the doors you walk into a long
building with a series of stalls on either side. A waist-high ornamental
guardrail keeps visitors from getting too close to the vertical bars of
the front of the stalls, but a good close view of the large variety of
antelopes is to be had. In the center of the west side the bars go clear

to the ceiling, and two eighteen-foot-tall doors are moved sideways to allow the reticulated giraffe to enjoy the much larger outside yard. Skylighting over all stalls provides excellent light for both the animals and the public.

Until he retired from the zoo in 1962, George Vierheller never lost an opportunity to tell people about the zoo, the animals, the animal shows, the birth of a baby animal, the acquisition of new animals, the development of new exhibits. In all media—newspapers, newsreels, news magazines, radio—and in personal conversation with all his friends, Vierheller's love for and pride in the zoo was transmitted to the people in the Saint Louis area, and they still support their zoo generously. Attendance gradually increased to approximately three million people per year. George invited visiting movie, stage, and concert stars to the zoo, and tipped off the press that the celebrities would be there. As a result he won acclaim and praise for his zoo from such notables as Lily Pons, Amos and Andy, Martin and Osa Johnson, Frank Buck, and even President Coolidge, who was not easily persuaded that the Saint Louis bear pits were artificial. Sunday mornings, the elite of Saint Louis and other cities came to the zoo to meet George Vierheller and be shown behind the scenes. George was a wonderful host; he loved people, and they in return loved him and his zoo. He had a small elevated platform built in back of the chimpanzee show arena, where he and his select friends could overlook the heads of the people seated below to watch the show. And if some of these friends were Betty Grable or Myrna Loy, they too could be seen by the visitors watching the show. George was never selfish with his friends.

I have been forever grateful to George Vierheller for giving me my start and helping me in my early years at the Saint Louis Zoo to find my way and know the best in zoo practices. He fostered many of my ideas and made possible their execution. At the same time he was building the zoo, he was building me as a zoo man. He became, eventually, the dean of American zoo directors and one of the most highly respected in the whole world.

When Osa and Martin Johnson came back from one of their many photographic trips in Africa, they brought some live animals for sale to zoos. George Vierheller arranged to buy a baby African elephant for the Saint Louis Zoo; the infant was to be delivered by the Johnsons themselves. They had also brought back from Africa one of their airplanes, a Sikorsky, which had been painted with giraffe designs. They agreed to fly the baby elephant with its Masai keeper to Saint Louis, with a refueling stop at Cincinnati. George Vierheller and his wife, Ida, went to Cincinnati by train, and at the appointed time boarded the aircraft and flew with the Johnsons and the baby elephant to Saint Louis. With two keepers, I met them at Lambert Airport, and the baby elephant was transported to the zoo in the back of a Buick sedan (with the seat removed, so the elephant would have firmer footing on the car floor). You can't imagine the double takes when the car stopped at a traffic light. The Johnsons called the baby elephant Toto Tembo, which is Swahili for "small elephant." The little fellow was only three feet high and still on a bottle. The Johnsons stayed in Saint Louis a few days to make sure the baby was well enough adjusted to be transferred to one of our regular animal keepers. On Sunday morning following the chimpanzee show, which the Johnsons had been watching from the side door, the audience out in front called for the Johnsons to come up on stage. Martin puffed excitedly on his long black cigar and said to Osa, "No, I can't go up there. You do it, Osa." And she did. She gave a fifteen-minute impromptu talk about their work in Africa and how they happened to catch these animals and bring them back to America. The Johnsons were charming and delightful people, and we were saddened two years later when Martin Johnson was killed in an airplane crash in southern California.

My years as curator of reptiles seemed to run together. It was such a fun job and so stimulating and absorbing that I couldn't imagine anything I would rather do. The work in the zoo was diversified, for I had to deal not only with animals, but also with people. Our annual snake hunts, which took us to such places as Arkansas, Louisiana, Florida, and Texas, were like paid vacations.

George Vierheller was like a second father to me. He was generous about asking me to do things around the zoo outside the normal duties of a curator of reptiles and introduced me to the many famous and influential visitors, asking me to show them special things of interest while he checked back at the office.

When I was twenty-eight years old I was married to Elise More of Saint Louis. Although the Depression of the 1930s brought hard times for many people, things did not cost much and my salary had increased to $3,300 per year. I always had an automobile, gas was cheap, apartments were inexpensive, and our food budget was about twenty-five dollars per week. I could pay the doctor bills, including the fees and hospital charges when our daughter Suzanne was born in 1937. Our cost of living did increase somewhat at that time, however, and I could foresee progressively increasing expenses as the baby grew older. It came as a welcome development, therefore, when in the summer of 1938 I was offered the directorship of the Zoological Gardens in Buffalo, New York. I did not like to leave Saint Louis, but besides giving me a larger salary the new job would heighten my position in the zoo field. The challenge of a more responsible position also fitted in with a credo I had long practiced—that of accepting challenges, extending myself, taking on a new responsibility when it was offered. Well, here was one that was offered, and I accepted. I talked it over with George Vierheller, and he seemed to understand the need for me to, as he said, "spread my wings." Despite this, there was a lump in my throat the day I gathered my personal possessions from my office and said goodbye to my zoo friends at Saint Louis.

9

The Buffalo
Years

The worst animal building I ever saw in my life was at the old Buffalo Zoological Gardens. However, the Works Progress Administration (WPA) had a long-standing commitment to the institution and was, in effect, building a whole new zoo. Fortunately, that old wooden building with its double-decker cages, dingy and dark and smelling to high heaven, was scheduled to be torn down. The only old building to remain was the elephant house, which was a one-animal building.

Big Frank, a full-grown Indian elephant with one tusk broken in the center, lived in that house. He was a really big animal. I had been curator at the Buffalo Zoo only a few months when in January of 1939, Big Frank died. Because of his size, I offered him to several museums around the country, but none of them wanted him. We were faced with the question of disposal. We had opened the doors of the building, and because it was a cold January, we thought his body would keep pretty well. But the day after his death it became evident that this was not the case. So when none of the museums would take him, I had to arrange for a soap factory in Buffalo to take Big Frank and, in return, supply the zoo with soap flakes for an indefinite period.

Officials of the Shrine Circus decided to replace Frank with another elephant. They found out about a bankrupt circus in Peru, Indiana, that had an elephant available. Sight unseen, and without consulting me, they bought her as a gift to the Buffalo Zoo. I learned about the donation only when a Shriner called to announce that the elephant would arrive the next day. When the big truck stopped in front of the elephant house, a sad-looking animal emerged, very thin, weak, and limping badly on one leg. The circus man with her said she had nothing to eat for weeks except dried corn stalks—not much of a diet for an elephant.

We fed her well and watched her closely. Soon we learned that when she lay down in the evening, she was too weak to get up again. I knew that if she lay too long on one side the reduced blood flow there would numb that side, and she would never get up. I alerted the night watchman to make half-hour checks in the elephant house, note the time the elephant lay down, and call me at home two hours later. I would then phone three keepers who lived nearby and all five of us, including the night watchman, would help the elephant up by pushing on her head until she could pull her front feet under her. She could rise the rest of the way by herself. This went on every night for about a month. Finally, she grew strong enough so that the night watchman and I could get her on her feet by ourselves, and in another couple of months she had regained her health and was able to get up without our help.

Her name was Bama. She was not a highly trained performer but had been used by the circus as a work animal, pushing wagons and helping to raise poles. When she got a little stronger and the weather was good, we would take her outside and let her walk around the grounds. We encouraged school groups to come to the zoo, and when they saw the elephant outside, the students would all congregate around her. Bama would be very careful with the children. They would come up and feel her body and legs and feed her. When she wanted to walk a little more, she would start moving her body gently from side to side and very carefully slide her feet forward, so that she wouldn't step on anyone. She turned out to be a very lovable, wonderful old girl.

The new zoo was in various stages of construction. The bear pits of natural limestone blocks had been nearly completed. They were moated and had dens in the back for the bears to retire to. The monkey island was completed and had had monkeys on it for one summer before I arrived. The main five-building zoo complex joined by vestibules in the form of a great horseshoe was under construction. The cages in the lion house had already been built. The wire and steel were there for the small-mammal house and the monkey house, but the interiors for the bird house and reptile house had not even been designed. The central section was three stories high and the whole building was constructed of local limestone. Some of this came from the zoo grounds when the pits were dug for the deer yards and the small-mammal area. The third floor contained a large room for the monkeys to live in during the winter months. In another area left undesigned I built a photographic darkroom.

When I first got to Buffalo the only place in the zoo where I could have office space was in the old, stinking building. The first time I entered that building, I threw up, so I spent very little time in my office. One of the happiest days of my life was when the construction for the big horseshoe-shaped main building had reached a point where we could move all the animals out of that old building and demolish it.

Old Doc Campbell had retired after a long career as curator of the Buffalo Zoo. Unfortunately, he had had very little money to work with, and very little new construction was attempted during his regime. I switched around the small-mammal house and monkey house and redesigned the cages. When I learned that the WPA organization had an art department I enlisted their help. Scenes of natural habitats were painted on the walls of the cages in the reptile house and a mural was done for the lion house. On a wall above the glass-front cage in the reptile house a frieze painted in monotone depicted some of those early reptiles, the dinosaurs. When I heard that the WPA had a woodcarver, I set him to work carving signs in the form of animals, pointing in the directions where the live animals lived. These signs were mounted on tall poles and placed in strategic locations around the zoo.

The bird house had a large central flight cage and fairly large glass-front cages around the walls. To dress up the exterior of the rather drab-appearing stone, the windows of the lion house and the monkey house were outlined in red and yellow sgraffito.

Eventually the entire complex was completed, and I had a nice double-roomed office on the second floor of the building. There was also an auditorium for meetings and a couple of rooms for use as classrooms or extra office space. A fine old rustic ornamental iron fence circled the eighteen-acre zoo. Just inside this ornamental fence were yards with moats separating the public from deer, elk, bison, camels, prong-horned antelopes, and moose. A large moated area with living trees on it formed a display area for Barbary sheep, tahr goats, and red foxes. Various moated stone-walled pits housed a variety of small animals such as prairie dogs, skunks, raccoons, young bears, opossums, penguins; and there was a very nice exhibit for beavers. In most zoos, beavers are nocturnal as they are in nature and are not visible to the public in the daytime. At the Buffalo Zoo, however, we fed them every afternoon at two fifteen, and they got so used to this they were always out on exhibit at feeding time and slightly before.

It was wonderful to be able to throw myself into my new work, but I felt the heavy responsibility of being in charge of the whole zoo. In addition, I felt handicapped by lack of funds. The Buffalo Zoo is a part of the park department, which is a part of the city government. This, of course, makes it a political entity. It was a nonessential service that each year had to compete for the tax dollar with the city's essential services. You know who got most of the dollar—the essential services. Parks, recreation facilities, and zoo didn't fare as well. Andrew Butler, a paint manufacturer in Buffalo who had taken the position of commissioner of parks, at one dollar a year, was the man who had engaged me. He had assured me I would have a salary increase at the end of the first year and that the zoo would have enough money for proper support. Having gotten involved, however, in the political machinations of city government, Butler resigned at the end of his first year. His successor was also a successful businessman, semiretired, but more of a politician. He was unable or unwilling to fight hard for the zoo. As a result, I did not get my raise and the zoo got a budget too small to maintain it. Our new reptile house was completed, but we had no money for keepers. I announced to the press that this building, although completed, lacked the necessary funds for maintenance and explained that the only way we could keep our reptiles alive was through volunteer help from a young man who was interested in reptiles and would spend as much time as he could after school and on weekends at the zoo.

As a result of that newspaper article, a number of young people volunteered their services. We selected a group of about twelve, and put them in the charge of Fred Myers, the only keeper for this three-keeper building. These boys and girls did an excellent job, not only cleaning the floor space of the building and the glass fronts of the cages, but actually taking care of the specimens as well. They enabled us to open the building to the public. The young volunteers formed their own club. In the space on the second floor, we gave each of them a cage for his or her own individual pet. We provided rats, mice, guinea pigs, rabbits, and various reptiles. Some brought in their own pets and took care of them at the zoo. These animals were never seen by the public, but it was a wonderful thing for the teenagers to assume responsibility and care for their own personal animals, in addition to caring for the zoo livestock.

My wife and young daughter, Suzanne, and I lived in an apartment directly across the street from the zoo. It was a very convenient location for me as I was often in the zoo after working hours and even at night. And when World War II started, I could save my gas coupons for more important travel than commuting. I accepted invitations to speak at all kinds of organizations in Buffalo in order to publicize the zoo. I spoke in churches, to Boy Scout troops, at civic organizations and schools; I even taught a course in herpetology at the Museum of Science. I became friendly with a number of people at the Museum of Science and eventually was elected to its board of directors. I was a frequent visitor to radio stations, telling the animal story, and the newspapers cooperated beautifully.

There were many attractive neighborhoods in Buffalo in those days, with elm trees arching cathedrallike over the broad streets. We were interested in music and the arts and entered into many civic activities in the community. Buffalo is a city of heavy snows and long winters. We enjoyed the winter activities—skiing, hiking in the snow, picnicking at places such as Chestnut Ridge Park. The park supplied firewood. We would dress warmly and go with a small group of friends to a fireplace area, where we would clean off the snow, build a fire, and have a barbecue. Darkness came early, and soon after dark the wild animals would arrive. Raccoons and skunks were particularly abundant. We always had a little food left over and we got well enough acquainted with the wild skunks to hand feed them. They would come up and put their front paws on our legs so we could drop food into their mouths. Some of our human friends were a little nervous when this first started, but unless a skunk is forced to do so, it is not likely to spray. On one occasion, we had eighteen skunks around us at one of these picnic areas. Even when we ran out of food, not once were we sprayed. As a matter of fact, I had twenty-two skunks complete with their scent glands in one of the moated areas at the Buffalo Zoo. Only once did a skunk spray, and that was when a keeper accidentally stepped on its foot.

Tom McLeary was animal foreman when I arrived in Buffalo and he continued to be animal foreman while I was curator. Talking one day about the importance of a relief keeper's following the procedures of the regular keeper, Tom told me of an experience he had had that well illustrates this point. Some years earlier, a white-tailed deer fawn had been born in the zoo and Tom had raised it. It turned out to be a beautiful buck, so outstanding that Tom kept it in a big yard all by itself. Each day Tom would go into the yard with a wheelbarrow, a rake, and a broom and clean the yard. This went on for two or three years and the deer would always come up to Tom and be friendly when he was inside the yard. One year, Tom took a month's vacation. He was replaced by a new keeper, appointed through the political system of the city of Buffalo. This man was afraid of the deer. Tom had accompanied him into the yard and had introduced him to the procedure and to the animal. But when he entered the deer yard alone, the new keeper took with him some loaves of bread. As the deer came up to him, he would throw a slice of bread way off and the deer would run after it. Then he would do a little sweeping up and when the deer came back, he would throw another slice of bread to keep the deer away. When Tom returned, the replacement keeper didn't say he had been feeding the deer in this manner. Tom went in as usual with his wheelbarrow, rake, and broom and started his work in the yard. The buck came over to Tom, who spoke to him. Tom assumed the buck was just being his normal, friendly self, and continued with his work.

The buck got behind Tom and, without the least warning, charged. He hit Tom squarely in the backside, knocking him flat, and gored him in the back with his antlers. Pulling his antlers out of Tom's body, the deer started at Tom's head. Lying on the ground, Tom realized he had to defend himself—it was early morning and there were no visitors in the zoo and no other keepers nearby. He also realized that he must grab the antlers, but must grab only on one side, so as to throw the deer off balance. If he grabbed both antlers the deer, who was stronger, would lift him off the ground. So he grabbed one antler and held it to the ground. The deer struggled, but with his antler held to the ground, he couldn't get at Tom. Tom then inched his way toward the gate, never letting go of the antler. After a terrible length of time, he worked himself clear across that large yard, got to the gate, let go of the antler, and with his free hand hit the deer as hard as he could in the nose. Then he rolled under the gate. Tom was in the hospital for three months. The antler had gone through his buttocks clear into his coelomic cavity and, of course, the wound became infected. He nearly died and spent nine months recuperating after he left the hospital.

Tom explained that he hadn't known the relief keeper had been paying off the deer. How different the story would have ended had there been good communication between those two.

Six cultural institutions in Buffalo—the public library, the Grosvenor Library, the Museum of Science, the Historical Society, the Albright Art Gallery, and the Buffalo Zoo—decided to join forces in a cooperative venture. The war was on and we chose the theme: The World We Must Know When the Boys Come Home. A series of exhibits depicted this in each of the six institutions. The libraries provided literature, the Albright supplied art objects, the Historical Society and the Museum of Science contributed artifacts, and the zoo provided wild animals. Our first exhibit, in the downtown public library, was on the South Pacific. A fine collection of books on the South Pacific from the two libraries was on display, artifacts from the various islands from the days before the white man got there came from the Museum of Science and the Historical Society, a Gauguin was lent by the Albright Art Gallery, showing a Tahitian beauty, and the zoo provided a tame and gentle live emu which was placed inside a picket fence in the center of the library.

Similar exhibits, each on a different subject, were held at each of the other institutions. The exhibit that drew the greatest amount of publicity was presented in the Grosvenor Library. The subject was the Near East. The zoo's contribution to that exhibit was a small cage of hamsters. A blessed event occurred during the second night, and when the news was released by the Grosvenor Library all the papers covered the story. Then a hamster escaped, and when the library released this news the papers printed humorous accounts. One quoted a librarian as saying, "I do hope it will be found in the zoological section." But the hamster's capture occurred not there but behind a book on the biblical history of the Near East.

The last exhibit was held at the zoo. The subject was Egypt, in recognition of the battles being fought at that time across North Africa. New and old were juxtaposed in this exhibit. It was fascinating to see a live Egyptian cobra and a sand viper casually crawling across a 4,000-year-old Egyptian necklace resting on a flat rock, then onto a somewhat tarnished metal mirror; and finally to watch the cobra flick its tongue at the hieroglyphics of an ancient funerary slab.

Each year, at budget time, I requested additional funds for the zoo as well as the salary increase that had been promised me. The budget was always cut, and I never got my raise. My annual budget requests included funds for such needs as repainting the wonderful wrought-iron fence around the grounds, which was rusting badly, and painting the windowsills in our new buildings, which had never been painted since the buildings were completed. None of these items was ever fitted into the budget. I pleaded with the city manager to visit the zoo and let me show him the need for the things I was requesting in the budget.

"Marlin," he said, "I have a firm rule in this office, and that is I never go into the field anywhere in the city of Buffalo."

"My gosh!" I said. "How can you possibly have such a rule? How do you know how to make a budget?"

"Marlin," he said, "I have only so much money to work with. If I come out to the zoo and you show me all these things, I'll know that you need them. But my job is to apportion the money for the city, and we never have enough. So I never go into the field to learn these problems first-hand."

About this same time, I saw one of those cut-up-dollar charts showing where the city tax money went. Amazingly, 62 percent of the money received from taxes went to pay the interest on the bonded indebtedness of the city. Even with the aid of the newly formed Buffalo Zoological Society, there was little hope of raising enough money to run the zoo properly.

Because of this impasse I was in a receptive mood when George Donahue, general superintendent of the Chicago Park District, asked me to become director of the Lincoln Park Zoo in the Windy City. I had known Floyd Young, the director of the zoo, for a number of years. He was about to retire. In addition, as assistant director of the Lincoln Park Zoo, I was to receive more than double my salary at Buffalo, with a still greater increase when I became director at the time of Floyd's retirement.

During my sixth year at the Buffalo Zoo, with Ward Soans, I had started an animal food business called K-9 Kitchen. This venture was to supplement my meager salary as zoo curator. Ward Soans was in the refrigeration business. He provided refrigerated showcases and freezers and other equipment, as well as enough money to get us started. My role in the partnership was the technical one of formulating nutritious feed for animals. Ward and I worked in the evenings preparing the food, packaging it, and putting it in the deep freeze. We employed a saleswoman to work in the store during the day. We also carried medicines, shampoos, soaps, leashes, and similar material for pets. The shop quickly became lucrative for us. The prognosis for such a business was very good indeed. If I had stayed with that, I would certainly have made a great deal more money than I did in the zoo. But the zoo was in my blood and I wanted to stay with my profession. Soon after I went to Chicago, Ward sold the business.

10

Escape of a Gorilla

To call on the director of the Lincoln Park Zoo was an embarrassing experience. The minute the blue-gray steel door opened from the vestibule into the office, he could be seen working at his desk. A visitor felt like an intruder, and there was no way to soften the blow; it was just one big room. Although a sofa and some straight chairs were provided for visitors, all conversations had to be conducted in the presence of the secretary, unless she was asked to leave, and then, of course, she was embarrassed. Sometimes, when the door was open, visitors come primarily to watch the animals would look in, and occasionally stroll over, to see who lived in cage number one of the primate house.

The arrangement was not efficient, but somehow the business of operating the zoo was accomplished from that office in Lincoln Park, facing Lake Michigan, with the lagoon and the outer drive in between. The north window of the office looked out on a great red-brick powerhouse with one of the tallest chimneys in Chicago rising above it. It supplied heat for all the buildings of that section of Lincoln Park. When I arrived at Lincoln Park as assistant director, an additional desk and chair were placed in the corner across from the secretary,

Miss Smith. I did some work in there, and Floyd Young and I had long conversations about the zoo and its operation. But most of my time for the first nine months was spent outside, getting acquainted with the zoo. I got to know the people who worked in the zoo, the repair and construction department of the park district that was responsible for the maintenance of the zoo, the electrical department housed in a section of the powerhouse, the landscape section that took care of the lawns and trees and the nearby greenhouse that supplied potted plants and flowers for our gardens. I spent a lot of time just walking around the zoo getting acquainted with the physical layout and with the animals.

I noticed that Tony Rausch, the keeper in charge of the bears, habitually entered the bears' area with his shovel and broom via the front gate near the public guardrail—when the bears were in that unit. I spoke to Floyd Young about this, and his reaction was, "Oh, I've spoken to Tony about that, but he's set in his ways." The method appeared to me unnecessarily dangerous as there were sliding gates at the entrance to the dens at the back wall where the bears could be transferred and the danger eliminated.

Soon after I was installed as the director of the Lincoln Park Zoo, I called Tony Rausch into my office and, in the presence of Richard Auer, the foreman, and Miss Smith, gave him a firm order never to enter any bear pit while the bear was present. I told him to catch the bear in the back den and then clean the pit. He argued a little and assured me, "Them bears is my friends and would never attack me." He offered to sign any kind of disclaimer paper necessary, if I would just not insist that he never go in with the bears.

I was firm in my decision, however, and reminded Rich Auer in Tony's presence that as foreman it was his responsibility to make sure my order was carried out. A few weeks went by and all reports and observations indicated that Tony had turned over a new leaf and was following the new directive.

Early one October morning, after Rich Auer had made his rounds of the bear pits and had again cautioned Tony not to go in with any of the bears, a lone visitor happened by the four small bear pits at the north end of the run and was horrified to see the Himalayan black bear biting and worrying the prone form of the moaning and feebly resisting Tony Rausch. The witness ran to the commissary building and told the keeper, who immediately alerted the office and then ran to help Tony. The office dispatched a man to bring a rifle from the lion house. But the Chicago Park District police had also been notified, and they arrived on the scene before the zoo rifle did. The police shot the Himalayan bear as he was trying to drag poor Tony through the door into his inside den. Tony was conscious but seriously injured. I arrived as he was being loaded onto a stretcher and into the ambulance. Tony said, "Mr. Perkins, I never thought he would do it." A team of surgeons labored over Tony for nearly eight hours, suturing and cleansing the nearly one hundred lacerations on his body. Despite blood transfusions and other supportive therapy, poor Tony died two days later.

In 1947 I was contacted by Otto Fuerbringer from *Time* magazine, who said that I had been chosen for inclusion in a cover story about the zoos of the United States. Two writers spent many hours with me at the zoo and elsewhere. They probed deeply, getting all kinds of background material, taking photographs, and going through the files at the zoo and at the offices of newspapers and magazines in Chicago, Saint Louis, and Buffalo. They talked to my friends and relatives in many locations, and Artzybasheff did a drawing of my head for the cover. When I knew for sure it was going to happen, I told the general superintendent of the Chicago Park District, George Donahue.

"Oh that's very nice," he said. "Very nice indeed. I'm glad to hear that. [Pause.] Did you say you're going to be on the *cover* of *Time* magazine?"

I said that was correct.

He said, "You mean your picture, your likeness is going to be on the front cover of the magazine?"

"Yes, Mr. Donahue," I said, "that's what I understand." He gave me a doubting look, but again said, "That's fine. That's wonderful! Great news."

When the article came out that summer the likeness was very good indeed. And as a result, I heard from friends I hadn't heard from in years, scattered all over the United States and a few abroad. Most of my colleagues in the zoo business were complimentary and thought the article was beneficial to zoos everywhere as it provided extensive publicity at the beginning of their summer season. I did, however, detect a certain coolness from a few colleagues, but then I suppose you can't please everyone.

The *Chicago Sun-Times* syndicate asked me to write a weekly column for distribution to newspapers throughout the country. I did so for about a year and a half, each column dealing with one species of animal. A photo of the animal illustrated each article.

Like many others, the Lincoln Park Zoo had its group of personality animals—animals so special that nearly everyone knew their pet names. Heinie the chimp was one of these. He had been born, as close as we could figure, in 1921; he lived a full and interesting life. Occasionally he entertained visitors with his version of the jungle stomp dance. His hair would fuzz up, his shoulders hunch forward, and he would start slapping the floor with his bare feet. He would move around the walls and kick the metal door with the heel of his foot, until the resounding metallic ring was audible clear over to the lion house. He would then work his way along the metal wall of the cage to the bars, grab hold of them and shake them, while glaring at the public outside. Finally, slapping the floor with his feet, he would repeat the metal-door routine, starting off slow and easy, building to a crescendo, and stopping as abruptly as he had started. He would then sit down and groom himself. Heinie had been acquired when he was two years old by a physician in Chicago, who had kept him until he grew too big to remain in a home any longer, then had given him to the Lincoln Park Zoo. Heinie was a gentle animal. He liked people and was particularly fond of his keeper, Eddie Robinson.

One hot summer day Eddie and foreman Rich Auer were standing on the porch of the north entrance of the primate house overlooking the thousands of people on the grounds. On the lawn in front a family had spread out blankets and a tablecloth and were picnicking. Suddenly, big Heinie came up to the group, put his arm around a lady who was sitting on the grass, looked into her face, and said *Ooo, ooo, ooo!* I think he wanted some food and was asking her for it. Rich and Eddie immediately ran toward the chimp, took hold of his hands, and said, "Come on Heinie, it's time to go." Heinie held their hands and walked to his cage; Eddie immediately put him into it, closed the door, and snapped the padlock. Heinie then suddenly realized he had been had. That was when he put on his greatest stomp dance. We never knew how Heinie got out of the cage, but the padlock was unfastened when Rich and Eddie returned there with him.

Heinie lived to be fifty-four years old. Even though some of his hair turned white and he took on the appearance of an old animal, he would occasionally put on a milder version of the famous Heinie stomp dance, even during the last few years of his life.

Heinie never learned to mate. He had been an only chimp for a good many years before he was given to the zoo. Then, even with an attractive female cagemate who wanted to start a family and gave Heinie all the encouragement normally necessary, he would instead relieve himself the only way he knew. We made it possible for him to see other primates copulating, hoping that he would get the idea. But his old habits were too firmly ingrained.

Even more famous and popular than Heinie was the gorilla Bushman, whom I inherited when I took over the directorship of Lincoln Park Zoo. Bushman had been purchased by Alfred Parker, then the director, from wild-animal dealer W. L. Buck, affectionately known as Pa Buck. (He was no relation to Frank Buck.) As Pa Buck, who never lost a chance to spin a good yarn, told the story, he was traveling through the French Cameroons in 1928 when he came to the native settlement of Yakadouma. There he learned that in a small village nearby there was a baby gorilla that had recently been captured by that tribe. Pa Buck hurried to the village, saw the baby gorilla, immediately fell in love with it, and started negotiations for purchase. After a time a financial impasse was reached. But Pa Buck, knowing the great respect the natives had for magic, was able to conclude the financial deal when he demonstrated his great feat of magic. He took out his false teeth, slipped them back in place again, wiped his hand across his face, and smiled. The gorilla was his.

Pa Buck named the little gorilla Bushman, a literal translation of what the natives called it. Soon afterward he returned by ship to the United States and docked at Philadelphia, where he offered the baby gorilla to the zoo. They refused because they had a male gorilla named Bamboo. Bushman was then offered to the National Zoological Park in Washington, D.C., which likewise rejected him because the reputation of gorillas for living in captivity was not good. Pa Buck then tried Lincoln Park, and Alfred Parker bought Bushman for $3,000 —a lot of money in those days. Pa Buck stayed in Chicago several weeks to transfer the affections of the animal from himself to his new keeper, Eddie Robinson.

Such was Pa Buck's tongue-in-cheek account of how he acquired Bushman. I myself often retold this entertaining yarn at dinner parties until, some years later when I was traveling through the Cameroons, I stumbled upon the truth. It was less colorful, but interesting in a different way.

From Dr. Albert Irving Good, the missionary in charge of the American Presbyterian mission station at Yaoundé, and his wife, I learned that Pa Buck had, in sober fact, bought Bushman from them. When the animal dealer came to Yaoundé he stayed at the mission because there was no hotel in the town. The Goods at that time were raising the male baby gorilla which they had bought in 1928 from some Africans who had captured him near the village of Yakadouma. The little gorilla was thin, weighing only fourteen pounds, and had a runny nose. Mrs. Good made a baby formula for him, starting with powdered milk to which vitamins were added. The Goods hired an African woman to take care of the baby ape like a surrogate mother; she carried him around, played with him, and gave him a chance to exercise.

As the baby grew he gained weight, ran and climbed, and became increasingly active. He was a healthy, happy animal. The Goods gave him free periods when he could have the run of the station area. He would frolic about, roll in the grass, and climb trees; but always he kept an eye on Dr. and Mrs. Good and frequently returned to climb into their arms for a hug and a pat on the back. The rapport between them was excellent. The Goods named the baby Bushman.

In a short time Bushman was playing with Pa Buck and they, too, became friends. Buck was so impressed with the young gorilla that he asked Dr. Good to sell him. At first Dr. Good declined. But just at that time the mission was building a stone church and needed money for stained-glass windows. Dr. Good realized that the young gorilla was growing rapidly and would eventually become too large to handle. An agreement was reached between Dr. Good and Pa Buck that Bushman would go back to the United States with Buck to be sold to a zoo, and half of the money would be returned to the mission. The mission's share of the $3,000 paid by Dr. Parker was used to buy stained-glass windows for the new church.

When Bushman arrived at Lincoln Park he weighed about twenty-eight pounds and was thought to be about two years old. He and Eddie became fast friends. In good weather Bushman had a play period outdoors with Eddie. They would wrestle, Bushman would climb small trees, and Eddie taught him how to play football. He loved this game because he could grab the ball and run with it and Eddie would chase him. When Eddie had the ball, Bushman liked to tackle with a one-armed grab at shoelace level. The people of Chicago took Bushman to their hearts. Scarcely a week passed without a picture of the gorilla in the newspaper or a story about his development or some exploit he was involved in.

At the end of the first year, Dr. Parker and Eddie thought Bushman ought to have a birthday party for the press. A nice birthday cake was arranged, the press was invited, Bushman threw the cake at the photographers, and some of them got good pictures.

As Bushman grew in stature, so did his popularity. Marks a foot apart were painted on the wall of his cage so when he stood up visitors could see how tall he was. A scale was built into the cage, so that when Bushman sat on his heavy metal chair his weight could be read by the people out front. Eventually he reached a weight of 570 pounds and a height of six feet and two inches. He had luxuriant black hair on his arms and body and shiny black skin on his face and ears.

One day, when Bushman weighed about 150 pounds, Eddie entered the cage carrying the gorilla's collar and rope for his romp outdoors. Bushman stood still while Eddie attached the collar, and then the pair moved to the outdoor section, south of the building, for a romp in the sunshine. Bushman thoroughly enjoyed that. They had been out only about five minutes when a keeper shouted to Eddie that a small monkey was loose in the building. Eddie thought he ought to help catch the escapee, so he took Bushman back to his cage. Bushman jumped up into the cage; Eddie followed, took off the collar, and started for the door. But Bushman was ahead of him. Clearly, Bushman was not going to let Eddie out the door; he wanted to go back outside and finish his romp. Eddie played with Bushman in the cage for a while, hoping this would change the gorilla's mood and make him forget his abbreviated outing. Ed made another try for the door, but again Bushman got there first, and his intention was clear. Eddie called a keeper to toss some grapes into the adjoining cage. Bushman dashed in, got the grapes, and was back at the door before it could be closed. They tried again with bananas, throwing them deeper into the adjacent cage, but Bushman, who had a reach of nine feet, held the sliding door open with one foot and was still able to reach the bananas. Again he came back into his cage before Eddie could make a dash for the door.

After two and a half hours of trying various diversionary tactics, one of the keepers produced a little squeaky toy. He made it squeak near Bushman and pretended he was going to throw it at the gorilla. Bushman ran to the far end of the cage next door and the keepers were ready. They closed the sliding door and Eddie finally got out. That was the last time Eddie took Bushman outside for a romp. He realized the gorilla had become too strong for him to control safely.

Eddie Robinson got to thinking about what it was like in the jungle and wondered where gorillas went when it rained. He said to himself, "By, golly, I don't believe they can get out of the rain; I think they must get wet. And maybe that's good for them. Maybe it helps to keep them clean!" He started giving Bushman shower baths, and Bushman loved them. The baths were so successful that a showerhead was installed above the cage, connected to a warm-water tap. Every morning, Bushman had his shower. This, of course, kept him clean and probably had much to do with his healthy appearance.

Many people have told me that, while driving through Lincoln Park, they would suddenly think of Bushman, park the car, walk in, and spend a half an hour with him. He was an active animal. When the wooden floor of his cage was wet, he loved to run and slide. He also liked to hit the quarter-inch-steel-plate door with his hand or with the heel of his foot; the resulting sound was like cannon fire, booming around the ceiling and walls of that great vaulted building.

As a special treat distinguished visitors were sometimes escorted behind the cages to see Bushman at close range, while Eddie or I fed him grapes or bananas. Bushman would come right up to the bars and put his lips between them, and we would feed him one grape at a time for as long as the bunch lasted. This gave the visitor time to see the size of the gorilla's fingers, which at the base were as big around as a man's wrist, and experience first hand his awesome size and power.

One Sunday morning the British consul general for Chicago, his wife, and a small group of friends came to the zoo, and I took them behind the cages in the primate house. I first showed off Dum Dum the gibbon, who would reach his long arms out of the cage to hug me; then the cherry-headed mangabey, who would come to the bars, sit down, and comb the hair on the back of my hand; then the other inhabitants of the long row of cages. At last we came to Bushman's cage. After a while the group began to talk and some turned their backs on Bushman. He didn't like that and quickly scooped up a handful of his own waste and threw it at them. It hit the cage bars and splattered. Most of it landed on the white suit of the British consul general. We scampered out from behind the cage into the kitchen. As we were attempting to clean up the consul general, and apologizing for our misbehaving gorilla, there occurred a lull in the conversation. The consul's charming wife spoke up and said, "Hum, anti-British, isn't he!"

We built Bushman a huge barred cage out of doors. He had injured one of his feet sliding on the wooden floor of his indoor cage, and we had been forced to substitute a polished concrete flooring. We thought the natural earth would be the best thing for him, and the new cage gave him the opportunity for the first time in many years of getting back on real earth. A heavy truck tire hung by a chain from the top of the outdoor cage. Bushman would swing that tire until it banged against the bars at the cage top. Sometimes he would twist the tire into a figure eight. Bushman spent many days during the summer months in that outdoor sunshine cage. But he was always fed in his inside cage at four o'clock in the afternoon, so he could be locked up for the night. He usually slept in the transfer passageway leading to the outdoor cage. The night animal keepers and those of us on the staff who sometimes tiptoed in to check on Bushman all knew that he snored.

One of the most exciting episodes in my career at the Lincoln Park Zoo was the day Bushman escaped. Eddie was breaking in a new keeper and showing him the routine of transferring Bushman to the cage next door. This was done so that his twin-cage compartment could be entered and properly cleaned. The two men worked together inside the two cages and when they had finished they stepped down into the passageway behind, closed the door, and slid open the transfer door to let Bushman back into his regular compartment. They then closed and locked the transfer door.

From there they went to the kitchen to prepare Bushman's food. They were cutting up carrots and celery at the sink when the new keeper suddenly said, "My Lord, there's Bushman!" Eddie thought it was a silly trick, but he did look around. There was Bushman in the kitchen, walking toward them. Eddie turned and walked toward Bushman and said, "Here, give me your hand, it's time to go for a walk." Bushman gave Eddie his hand and together they went to the back of the cage, where the door was open. Eddie said, "Okay boy, up you go." Bushman started into the cage, then changed his mind, turned, and bit Eddie on the arm. Then the gorilla turned and ran down the corridor. Bleeding from the deep puncture wounds in his arm, Eddie told the new keeper to spread the alarm and order all doors secured. By the time the two men reported to my office, the building had been evacuated and all the doors were closed and locked.

We alerted the park district police, who came in considerable force with a variety of weapons. Bushman was confined to the kitchen area and the long passageway behind the cages leading to the office door. He also could go downstairs to the basement and up the stairs on the opposite side of the cages to the room comparable to the kitchen on that side of the building. While Bushman was investigating the basement, Eddie Robinson slipped in through the office and closed another door in the corridor, thus confining the animal to the keepers' passageway behind his own cages. Bushman was spending most of his time in the kitchen and in the passageway behind his twin-compartment suite. The back door of one of the twin cages was wide open. We decided to try to lure Bushman with food. But the fresh fruit was kept in refrigerators in the basement. Lear Grimmer, assistant director, volunteered to fetch some grapes. He sneaked down the west side of the building to the basement, crossed to a refrigerator, softly opened its door, reached in to get a handful of grapes, looked up at the stairway, and saw Bushman standing there. Grimmer let the refrigerator door close and made a wild dash for the stairs on the other side of the building. He raced up the stairs, through the doors, and out into the public spaces of that building. He had the grapes. Maneuvering from the front of the cage, we put the grapes in the north compartment of Bushman's suite. We had previously placed a stick against the sliding door that led into the other compartment so the door could be pushed shut if the lure worked. Bushman saw the grapes go into the compartment, looked in through the sliding door, sized up the situation, entered the north compartment, held the sliding door open with one foot, stretched out one arm, picked up the grapes, went back into the south compartment, out the other door, and back into the kitchen, eating the grapes. We tried again with freshly cut mangoes; again he outsmarted us.

Knowing that Bushman was afraid of a baby alligator, I sent to the reptile house for one. As it was a warm summer day, the window to the kitchen was open and there was a heavy diamond-mesh screen on that window. I wanted to try to scare Bushman into his cage so we could slip into the kitchen and close the doors leading to the passageway behind his suite. I tied a string around the baby alligator behind his front legs and attached the other end to a bamboo pole. The keeper was told to slide the alligator through the diamond mesh of the screen and extend it as far and as fast as he could toward Bushman. I was watching through a crack in the double doors leading to the public space. Eddie Robinson and another monkey-house keeper were by my side. Bushman was in the kitchen and we could see him through the crack in the door. We saw the alligator being put through the diamond mesh of the window. Bushman looked up just as the long pole was thrust in his direction. He let out a piercing scream and ran into his cage. We opened the door, dashed into the kitchen, slammed shut the two doors leading behind the gorilla's suite, and barricaded them with benches and boards. We had now confined Bushman to the area of his suite, but he could still come out of his cage into the passageway behind.

I then sent to the reptile house for a snake small enough to crawl underneath the north doors leading to the passageway behind Bushman's suite. A medium-sized garter snake was brought over. Lear Grimmer was hiding in front of the cage, between the bars and the glass partition wall, his hand on the stick that would close the sliding door once Bushman got into the north compartment. I slid the little snake under the door. I heard Bushman climb up into his cage, and then I heard the wonderful sound of the sliding door being moved to the closed position. The bar that pushed the door open automatically dropped into position, locking that door. At last Bushman was back in his compartment!

My next action was to examine the padlock on the door to the south compartment whose failure to hold had enabled Bushman to get out of the cage to begin with. The lock worked perfectly. It must have been a human error, but I was so happy to have Bushman safely confined that I couldn't bring myself to scold anyone.

When the Ringling Brothers Circus was playing at Soldiers Field in Chicago I received a call from Henry Ringling North saying he'd like to come to the zoo on Sunday morning. I was delighted to meet him and to show him around. His first question was about Bushman, so I said, "Well, we can go see him right away if you'd like." Bushman was in his new outdoor cage. North and I and North's chauffeur went around to the outside cage. North went inside the guardrail to get a close look; the chauffeur stayed outside. Bushman was on all fours and fairly close to the bars. North looked him over carefully. He then turned to his chauffeur and said, "Fred, standing the way he is now, he may be just a little bit taller." Bushman walked over to the other side of the cage and stood with one arm in his tire. Again North turned to Fred and said, "Standing upright, now maybe he is just a little bit taller." He looked Bushman over, from his head to the tips of his toes, you might say, and in every position, he had to admit that Bushman was a little bigger than the huge circus gorilla Gargantua. That was all North wanted to see at the zoo; he wasn't interested in anything else and soon departed.

Years after he retired from the animal-dealing business and moved to Florida, old Pa Buck used to come up to Chicago in the summer and visit his old friend Bushman. In his later years he had to use a cane, but he made the trip and enjoyed spending a couple of days around the zoo. Most of the time he was in front of Bushman's cage.

Another animal dealer who visited the Lincoln Park Zoo was Snake King from Brownsville, Texas. He and I were walking around the grounds one day when we met a reporter from one of the local newspapers. I introduced King, and the reporter asked him what he thought of Bushman. Without any hesitation, Snake King said, "That Bushman is the most valuable zoo animal in the world!" The reporter wanted to know what kind of value he was talking about. King said, "He's worth every penny of a hundred thousand dollars!" And at a meeting of the American Association of Zoological Parks and Aquariums held in Saint Louis Bushman was acclaimed the most valuable animal in any American zoo by my colleague and friend Robert Dean, director of the Brookfield Zoo. This made good copy for the local news media and raised Bushman's status even higher among his host of friends in Chicago.

11

To West Africa for Gorillas

One of Bushman's greatest fans was a successful Chicago manufacturer, Irvin Young. In his early years, Young had been a missionary with the American Presbyterian mission in the French Cameroons. He had been to Yakadouma, the village near where Bushman was captured, and he still spoke some of the native language. During one of his many trips to the zoo, he spoke to me about Bushman's age and asked what we were going to do about a replacement for him in the future. I'd been thinking the same thing. Nobody at that time knew how long gorillas lived. Based on information about chimpanzees and orangutans, we thought gorillas might live to be forty-five or fifty years old. But Bushman had injured a foot; it had not healed well, and one of his toes was missing. He looked as though he was slowing down somewhat. It was time to think about a possible replacement.

Irvin Young thought that, through his continuing connection with the American mission in the Cameroons, he could help us acquire two or three young gorillas. I would procure the required government permits, arrange transportation, and find the money necessary to complete the trip.

I discussed the plan first with my superior, Red Weiner, director of special services of the Chicago Park District, and with George Donahue, the general superintendent. Both agreed that the idea was good. I found that a colleague, Professor Urbain, the director of the two zoos in Paris, was also consultant to the French government in relation to France's African territories, and permits for the capture and exportation of living gorillas were easily arranged. Professor Urbain also offered to rest the animals at the Jardin des Plantes zoo on their way to America from West Africa. TWA agreed to allow me to fly in a cargo plane from Paris to Chicago so that I could take care of the animals en route.

Toward the end of August 1948, with all my arrangements completed and with $3,000 allocated by the Chicago Park District for the journey, I departed from Chicago on a TWA DC-3 to New York, where I transferred to an overseas flight for London. There I had a four-day wait for a BOAC flight to Lagos. My Marble Arch hotel was only a short distance from the Regents Park Zoo. There I met Professor Hindle, director of the zoo, who took it upon himself to make sure I was properly fed and shown the sights of London. In 1948 English life was still austere as a result of the war. Food, clothing, soap, and other commodities were in short supply. Professor Hindle took me to his club for dinner, and this was a delightful occasion. The following day we lunched at another club and dined at still a third, which turned out to be the Service Club for members of the armed forces. Professor Hindle explained that the clubs were the only places where you could get a good meal in London in those days.

Finally I left for Lagos, and after an overnight there took an Air France DC-3 for the flight down the coast to the seaport town of Doala in the French Cameroons. There I was met by the business manager of the American Presbyterian mission. He took me to mission headquarters, a two-story brick-and-concrete house originally built by the Germans. The plane for Yaoundé did not leave for three days, and that gave me an opportunity to rest from my long trip and to see Doala. The second day I asked about the possibility of buying African souvenirs, so my host escorted me to the native market, where I bought some interesting blue-and-white cotton cloth which the natives used for dresses and shirts. I didn't see any wood carvings, so I asked about them. "I have a wood carver right here," my host said. "He's carving a bunch of little elephants for me. I'm going to take them back to all the people who have been contributors to our efforts out here. I have a whole trunk full of them. Why don't you help yourself? I'll just charge you the cost of manufacturing."

The wood carver was working under a little thatched roof to protect him from the sun. He was making elephants of various sizes, from about one inch to four inches long. They were carved from ebony. I picked up a couple and after looking at them for a minute realized that these were Indian, not African, elephants. "How," I asked my host, "does it happen that he's carving Asiatic rather than African elephants?"

"Oh, is there a difference?" he asked.

"Of course there is!"

"Well," he said, "that fellow has never seen an elephant. I employ him because he knows how to carve. I gave him a picture of an elephant and said, 'Make me a whole bunch of these.' "

I brought some home anyhow, because I wanted some kind of wood carvings. They have been conversation pieces ever since.

Word had arrived before I left Chicago that two young gorillas had been captured and were waiting for me in Yaoundé. In Doala, I heard that a third had been obtained. When my DC-3 landed on the graveled runway at Yaoundé, I was met by Dr. Good, the missionary in charge. He took me up to the mission house where he and his wife lived. Mrs. Good was a charming woman and sympathetic to animals, especially gorillas.

"Come on," Dr. Good said, "we'll show you where your animals are." The Goods led me to a building where a game scout from the conservation department of the Cameroons was in charge of my three gorillas. The animals were in makeshift box cages, but were well cared for. Everything was clean, and two of the animals looked fine. But the largest one, a male, had sores on his face. Dr. Good explained that the animal had been infected by yaws when he was newly captured by a family in a small village. Treatment with salvarsan had already been started, and would, they hoped, cure the condition.

The next largest gorilla was a female; the smallest, another male. The animals were being fed the local version of French bread, palm nuts, bananas, oranges, cassavas, peanuts, and mangoes. They ate well, and each had a glass of powdered milk before retiring for the night.

Dr. Good also told me about Bushman's babyhood and about how he had arranged with Pa Buck to take the gorilla to the United States, sell him to a zoo, and divide the proceeds with the mission.

"Let me show you the stained-glass windows that Bushman paid for," Dr. Good said. We walked along a path that led down from the hill to a stone church, and as we walked down the church's center aisle we looked up and saw the beautiful stained-glass windows, depicting the Nativity.

That night dinner was announced by the sound of a talking drum—a vibrant, throaty resonance, and one of the most interesting drum sounds I've ever heard. We were awakened the next morning by the same drum, and after breakfast I went out to examine it. This drum was a log about five feet long and thirty inches in diameter, with a slit about three inches wide in the top which extended nearly the length of the log. The inside was hollowed out through the slit. It rested on two smaller logs that were notched to keep it from rolling off. On each side of the slit was a rounded lip, one larger than the other. Dr. Good explained that the larger of the two made the male sound when struck with a drumstick, and the smaller the female sound. As the languages of this part of Africa are tonal, it is possible to transmit messages over long distances with a drum like this and to be answered by a similar drum a long distance away. Messages are sent from village to village and relayed on and on with drums. During my three-week stay at the mission, I heard the drums every day and frequently an answering drum far off on a distant mountainside.

I visited the native market with the game scout and found it a fascinating place. Many different kinds of fruits and vegetables were sold, including tomatoes, sugarcane, corn, squash, and pumpkin. Also available were colored cloths decorated with African designs but actually printed in Belgium. In the drugstore were animal skins, skulls, claws, and teeth, as well as turtle shells and various kinds of plants. At one end of the market was a bakery where the game scout always picked up a few loaves of French bread. The flour was infested with weevils, and because the insects were so hard to remove, the loaves were baked with the weevils included. At least there was a good crunchy crust on the outside.

I had brought with me a snake stick and some snake bags, hoping to find time for collecting. Each day after we had put the gorillas to bed at five o'clock, I would put on my boots, get my snake stick and bags, walk down the hill to a little stream, and work my way up through the foliage on the stream banks. The going was hard; this was tropical country and cat claw vines grew all around, devil's walking sticks barred my way, and the foliage was so thick that visibility was poor. I hunted quite a few days but never once saw a snake. After I had been at the mission about a week, however, I was awakened in the night by Dr. Good, who came into my room in his nightshirt, saying, "Come quickly, we have a snake in our room!" Sure enough, they had been visited by a nonpoisonous file snake, which I was able to catch and add to my collection.

In the market I was able to buy a female mandril baboon, a green monkey, and a baby Guinea baboon. My collection was further augmented by delivery of a fourth young gorilla, which had been captured nearby. He was a tiny baby, weighing only about ten pounds. He was skinny and dirty, but otherwise seemed to be in good health, although there was a little mucus in his nose. He was so young he needed somebody to cling to, and I provided that refuge during most of the day. The game scout carried him around when I had other things to do. I had brought with me baby bottles and nipples, powdered milk, multivitamins, and trace minerals, so it was not difficult to develop a formula for the baby gorilla. He took to it readily and received a bottle three times a day. Because of his slight cold we didn't bathe him at first, but when the cold was gone, the game scout heated some water in a tin can and he gave the baby a bath. Of course, he objected like mad and screamed and wriggled, but the scout was persistent and patient and gentle, and eventually, the ordeal over, the baby enjoyed a rubdown with a towel.

For the trip home I had boxes made for each animal, constructed of lightweight but strong wood. They were solid boxes with holes drilled in them for ventilation. A Hausaman came by one day with a big bundle on his head. He was selling blankets and leather hassock material. I bought three blankets, one for each of the two largest gorilla boxes, and the third for the two smaller gorillas who were traveling in a single box. The day we left Yaoundé, the first four seats of the DC-3 were filled with cargo—our animals in their boxes. We stopped in Doala for a short time and then flew on to Lagos. This was an overnight stop, but there was a rest house at the airport that passengers could use, and I took my animals in the room with me. They had a good meal and a good night's rest, and so did I.

The next day I presented a letter from the head of Air France requesting the plane's captain to allow me to put my cargo of gorillas in the forward compartment of the aircraft, rather than the belly. The captain shook his head and said, "No, this is not possible!" I argued with him. He spoke perfectly good English, but the minute I protested, he shrugged his shoulders and insisted he spoke only French. The Air France representative at Lagos took my side and indicated that the captain should heed this important letter from a top official of the airline. But the captain refused to fly the plane with the animals in the forward compartment. Frustrated and disappointed, I had to yield and allow my gorillas to be stowed in the plane's belly. The captain promised to fly at a low altitude for the comfort of the animals.

Our first stop from Lagos was Kano. The minute the DC-4 was on the ground a bus arrived to take the passengers to the terminal for food. I said, "No. I must feed the animals first. If there's time left I'll eat."

The crew took the animals' boxes out of the belly compartment and put them on the pavement. I examined each box, fed all the animals, gave them milk and water, and made sure that their temperatures were normal. By the time the animals had finished eating and I had cleaned their cages, the other passengers had returned, ready for the long flight over the Sahara. They had brought me a box lunch to eat on the plane. The animals went back in the belly compartment, and we flew on through the night to Tripoli. The same thing happened when we landed in Tripoli: I stayed with the animals and fed them, but had no time for breakfast at the terminal building. Again, my fellow passengers brought me a box breakfast.

An hour or so before landing in Paris the plane's captain, who had changed into his blue uniform for the north, rather than the khaki shorts he was wearing in Africa, came to my seat at the front of the plane, bowed politely, and asked if I would join him in a glass of champagne. I accepted. He was once again speaking English, and he tried to justify his refusal to allow the animals in the cabin. He said the smell would have been objectionable, but I think it was more his nose that was being affected than those of the passengers.

The captain had radioed ahead to alert the Paris zoo to our arrival. A truck from the zoo met us at the airport and transported all the animals to a building at the Jardin des Plantes. There the zoo attendants had food for my animals—lovely French bread, apples, oranges, bananas, grapes, pears, and peaches. I opened the door of one of the gorilla's cages, put some food in front of the partially opened door, and waited to observe the reaction. First I offered grapes, but the gorilla flicked them away. He had never before seen a grape. I then put a grape in my mouth and smacked my lips to let him know how good it was. Once he tried it, he liked it. The same thing happened with apples, pears, and peaches. Well, I thought, there certainly would be no trouble with bread; the gorillas had been used to eating bread in Africa. But this was all-white bread. It didn't have any weevils in it. And none of the gorillas would eat it. They loved the bananas and the oranges, which they knew, and they drank their milk and their water. I always feel it is not a good idea to overfeed animals when they're traveling, and these had had a long trip in the belly of the Air France plane.

Our TWA flight left the following day and I was given a section marked off with canvas partition walls and a big net to go over the boxes at takeoff. There was an auxiliary seat for me just back of the cockpit. Once in the air, I could take the net off the cages when it was necessary to feed and care for the animals. All the crew members were interested. When it came time to feed the baby male gorilla, I took him into the cockpit to let the fellows see him nursing a bottle. I had him wrapped in a piece of blanket and he lay in my arms drinking until he was finished. Then he raised his head and stared around at all the instruments. He reached out to touch the gold braid on the captain's shoulder. The captain offered to hold him, and the baby curled right up in his arms and then sat up in his lap. The captain was holding the wheel with his free hand, and when the baby gorilla reached out, the captain put the tiny hands firmly on the wheel, took his own hand away, and for a moment or two the baby gorilla was flying the DC-4 over the North Atlantic—long enough at least for me to snap a picture.

As we approached a weather ship somewhere in the North Atlantic, we dropped down where they could see us, for radio broadcasts had spread the news of the unusual cargo on this flight to Chicago. When we came in over New England word went out to a Boston station to inform my colleagues at the annual meeting of the American Association of Zoological Parks and Acquariums that at that moment Marlin Perkins, four baby gorillas, and a few other animals were flying in from West Africa. An announcement was made at the meeting. We landed in New York and transshipped on to Chicago, where we landed at Midway Airport. The press met the flight and took pictures of us unloading the animals from the plane and later transferring them to the primate house at the zoo. The keepers had cages all ready for the animals, we fed them and then tucked them in for the night. By that time I was pretty tired, too. My time clock was out of kilter, and I was glad to get home.

12

Personalities and Progress at Lincoln Park

.

We had a contest to name three of the gorillas. I insisted on calling the largest one Irvin Young after the benefactor who had made it possible for us to go to Africa and get them. As a result of the naming contest, the female was called Lotus, the baby Sinbad, and the third Rajah. We introduced the animals to Bushman, but only at a safe distance; we didn't want anything to happen to those precious babies and we didn't know what Bushman's reaction might be. Sinbad weighed eleven pounds when he arrived at the zoo, and the people who were frequent visitors were able to watch his growth and development. One shot of penicillin cleared up the yaws on Irvin Young's face and he became a fine-looking animal.

Just as zoos have personality animals they also have personality keepers, and Lyman Carpenter was one of these. He had friends all over Chicago, and many people came to the zoo to see him. He had a fine rapport with animals as well. And in addition, he was an artist and craftsman who could build almost anything. Lyman produced some really good ideas for the Lincoln Park Zoo. One of these was to provide a children's zoo. With plywood and paint, he put together an attractive small zoo. This was so successful that eventually the present children's zoo at Lincoln Park was designed and constructed. And it probably never would have been without Lyman Carpenter.

Another of Lyman's ideas was a zoo nursery. In most zoos, a newborn animal which needed constant, twenty-four-hour care was taken home at night by a keeper or staff member. Lyman thought it would be better to bottle feed the infants at the zoo. We tried it, and the project worked so well that I thought the public ought to be allowed to see some of these activities. A zoo nursery was built in the lion house. The nursery had glass windows and glass-front cages inside. We acquired incubators to keep constant, even temperatures and humidity for the babies who required these. All formulas were posted; the babies were weighed and their weights displayed. There was space on the nursery floor where young tigers, lions, jaguars, and monkeys could play and exercise. All this was so successful that a much larger and better zoo nursery was built as an adjunct to the children's zoo. There was always a happy group of visitors who were delighted to see these babies given expert care and attention.

Lyman acquired much useful information about caring for baby animals. For example, he noticed that mother animals often lick their babies, and that the licking triggers urination and defecation by the babies. Lyman put this information to use when a litter of orphan raccoons was given to the zoo by a person who had had them for a little more than a day. The kits were brought in late in the afternoon and Lyman was soon looking after them. He noticed that their stomachs were badly distended. He then remembered he had seen mother animals licking their babies. He moistened a finger and simulated the action of the mother's tongue. Shortly the kits started to urinate and defecate, and before long the distention was relieved. The term "piddling" was invented to describe this action, and all personnel in the children's zoo and in the nursery were instructed in this technique.

The animals in the nursery proved a great attraction, and as soon as they grew old enough, they graduated to the children's zoo. They were already tame and gentle and imprinted to human beings, although, we always tried to minimize that imprinting and give them members of their own kind as playmates and companions.

Until I became director of the Lincoln Park Zoo, the staff consisted of the director, the foreman, and the animal keepers. When Richard Auer retired I instituted the curatorial system and made George Irving our curator of birds and eventually general curator. Through the civil-service department of the Chicago Park District, I devised the position of zoologist, and a job description was written for the civil-service department. Fred Myers, who had been with me at the zoo in Buffalo, came to Lincoln Park as an animal keeper and later became a zoologist.

Fred Myers started the zoo's answer shop and ran it for some time. He tabulated the questions asked and found that many people were interested in the same subjects. The most often asked question turned out to be, "What is the gestation period of an elephant?" The answer is from eighteen to twenty-one months, depending upon whether the baby is a boy or a girl. Boys take a little longer.

In upgrading the zoo, I wanted to devise new labels for all the exhibits. For years, I had written labels for the Saint Louis Zoo and the Buffalo Zoo. The standard label gave the animal's common name, its scientific name, its habitat and range, and included a forty- to fifty-word description of the animal's natural history. The thought behind this was that if you put all the labels together, they would make the equivalent of a condensed book on natural history. People could get quite a bit of information about the animals by visiting the zoo. But they didn't. When I walked around in front of the cages and listened to the talk and watched the reading habits, I soon found that most people didn't read the labels at all. They just read the common name. I was interested in education, and I wanted to make those labels more appealing. In my exasperation one morning, I said to Fred Myers, "Fred, I wish you had never seen a standard zoo label. I wish you could erase the concept we have always had from your mind and start completely fresh with something new and different. I think we ought to take a tip from *Time* magazine and condense as much as we possibly can. Maybe we don't even need complete sentences. Perhaps we should eliminate all the descriptive adjectives. Maybe we should use arabic numerals instead of writing them out. It will give more white space on the sign and make it more quickly readable. Fred, go away and think about that and come back with some kind of concept."

In about a week, Fred came back with a tabulated sign. It started out with the common name of the animal; next to that was a heading, Other Names. Frequently, an animal such as a cougar has alternate names: mountain lion, puma, panther, painter, catamount, El Leon. Then came the heading, Gestation; this was included because of the interest expressed in the gestation of an elephant: The next line told the animal's size—height, length, and other pertinent measurements. Next was weight. Then came range and type of country the animal lived in. Then voice. Then life expectancy. And finally the scientific name. A couple more lines gave other pertinent information, such as number of young per litter or some quality useful to man. We conveyed more information in that tabulated sign than we had ever given in the old paragraph-type labels. The new signs were easier to read and many more people read them. Important animals such as elephants had large signs describing their personal history and detailing each animal's exploits. Large signs for animals like Bushman told where he had come from and when, his weight on arrival, how fast he grew, amusements he liked (how he played football with Eddie Robinson), and perhaps a condensed version of the natural history of the species. Such large, specialized signs are needed in zoos. If they are put together in an interesting way, they will be read by the public.

13

On the Air
with "Zooparade"

The lifeblood of a zoo is publicity and promotion; so in 1945 when I met a director of the experimental television station in Chicago, WBKB, and was invited to come to the studio with some animals from the zoo and give a talk about them, I jumped at the chance. Television seemed the perfect medium for the Lincoln Park Zoo. At the station in downtown Chicago, I saw my first television camera and studio and learned that there were 300 receivers out there in Chicagoland. In those days, videotape had not yet been perfected, so all TV shows were live, happening at that very moment. Film inserts were used, but only for short segments.

I asked how long my talk should be and the director said, "Oh, just talk until you run out of steam and then we'll have a chalk talk to follow that." I had brought along a keeper to take the animals out of their carrying cages and their sacks, to hand them to me, and to receive back the ones I had finished talking about.

On that first occasion, I took along a bullfrog and talked about its adaptations. The frog sat on the desk top, his throat pulsating and showing clearly that his throat was pumping air into his lungs. I called attention to this and explained that the muscles of the throat were designed for pumping air and, therefore, were not very good as swallowing organs and probably accounted for the frog's short neck.

I then described another adaptation that enables a frog to swallow the things he eats. He secures food by flipping out his long, sticky tongue, which is attached at the front of his mouth. The tongue flies out, grabs the insect or worm, and flips it back into the mouth. Then the frog retracts his eyes, which are quite large and positioned on top of his head; the muscles pull the eyes down through two openings in the skull, and that pushes on the roof of the mouth, starting the food down the short journey to the stomach. This was demonstrated to the audience by moving the camera in very close. I would touch the frog on the top of his eye and he immediately retracted it.

I did some fifteen half-hour shows at WBKB between 1945 and 1947. One day I was at the Museum of Science and Industry on the south side of Chicago and noticed on the street a WBKB remote-control bus. I stopped to talk to some of the crew, who said they were there to do a pickup from the museum, which was to be broadcast over WBKB. I went right back to my friend the director and suggested that in future I would like the bus to come to the zoo so we could do the show direct from there. An in-zoo show would have much greater variety than we could manage by carting small animals to the studio. The director inquired, but reported that Bill Eddy, who was in charge of the station, had vetoed this idea as too expensive. "Why," I asked, "if it's not too expensive to take the bus to the Museum of Science and Industry, is it too expensive to take it to the zoo?" I told the director I would no longer bring zoo animals to the studio, but would be glad to cooperate if WBKB wanted to bring the mobile unit up to the zoo.

While doing the shows at WBKB I had met Reinald Werrenrath, Jr., and Don Meier, two men who had returned to civilian life from military service in 1945. They had seen my presentations at the studio and I had had interesting talks with them about the future of television. In the spring of 1949, "Werry" came to the zoo to check with me about doing a program on the NBC-TV station that had recently begun operating. The coaxial cable from Chicago to New York had just been completed, and the station in Chicago wanted to show New York what kind of programs it could develop. A program directly from the zoo was proposed, using the remote-control unit. Along with the equipment they sent a newsman, Jim Hurlbut, to open and close the show and help out in case I got tongue-tied and didn't know what to say. He would ask a question and I would go on from there.

Following that program, in which we showed our famous gorilla, Bushman, along with other primates, Werry proposed we continue these programs on a weekly basis for the Chicago audience. This seemed like good publicity for the Lincoln Park Zoo, so I agreed. Each week that late spring and summer on Sunday afternoon at five o'clock we went on the air directly from the zoo. We called the program "Visit to the Lincoln Park Zoo." When the Korean War started, Werry was called back into the navy, and Don Meier became producer-director.

I thought the show would run through the summer until perhaps Labor Day and then when the kids went back to school would be discontinued. But we kept on going, and by October it was evident that NBC Chicago wanted to continue through the winter. We cleared an area in the basement of the reptile house that was large enough to hold the cameras and equipment. That was our studio in the zoo from which we did some of the programs.

We worked out of every part of the zoo and presented programs about lions, buffaloes, bears, the ducks on the lakes, the exotic birds in the zoo rookery. When it became evident that the show would continue through the winter, I suggested the name "Zooparade." The TV people agreed this was a better title than "Visit to the Lincoln Park Zoo," and it became the official name of our program.

In the spring of 1950, I was startled to learn that the program had been sold to Jewel Food Stores in Chicago and was to be a commercial show. I suggested I might be paid for my services; they agreed, and I signed a contract with NBC. A few weeks later "Zooparade" became a network program carried by about twenty-eight stations when the Quaker Oats Company bought the network portion of "Zooparade" for Ken-L Ration dog food. They could not get the Chicago outlet because it was already sold to Jewel Food Stores. For almost a year we did two shows each Sunday afternoon. The network show went on at five, and at six we did another show for the Chicago viewers.

We developed a script of sorts for the "Zooparade" program; we had rehearsals; and I tried to say the same things on the six o'clock local shows that I had said on the network. But the two shows were never quite the same because most of what happened on "Zooparade" was extemporaneous. The script was a general one that outlined the subject matter, but it was up to me to put the show together and fill in the outline. If I had been asked to do a television program ten years earlier, I could not have done it. I just didn't have enough knowledge about zoology and animals.

On "Zooparade" we worked out a system with our animal keepers. They brought their animals to the basement of the reptile house, cared for them there, and took them back and reintroduced them to the cages after the show. They were paid by NBC for this, as were George Irving, the general foreman, and Lear Grimmer, who was curator for a while and then became, in effect, my assistant at the zoo. Irving and Grimmer alternated each Sunday; they were responsible for seeing that the requisite animals were on location and that all props were available and in good working order. "Zooparade" was possible at the Lincoln Park Zoo because I had encouraged the keepers to make pets of their animals whenever possible. The animals were used to having people touch them, or pick them up if they were small, and were well acquainted with their keepers. If I had tried to develop a training program to teach keepers how to handle and move animals, I couldn't have found a better medium than "Zooparade."

"Zooparade" was good for the Lincoln Park Zoo in several ways. It was a training program not only for the animal keepers, but for me and for the curators. It brought an enormous amount of publicity to the zoo, to Chicago, and to animals. It gave me an added income, as it did the other zoo men who worked on the show. And it also added considerably to the income of the Chicago Park District, which eventually entered into a long-term contract with NBC. This made it necessary for NBC to come back to me and renegotiate a long-term contract for my services. From this I made more than twice as much as I did as director of the zoo.

Each week featured a Monday morning conference at the zoo with George Irving or Lear Grimmer, whichever was going to work the following Sunday. Don Meier and Harry Mahl, chief technician from NBC, attended. We outlined the subject for the following Sunday, talked about the story line, discussed how we could visually depict it and whether some film was needed. Don would schedule Marshall Head or some other NBC photographer to shoot whatever movies were needed, and these were inserted into the program while it was on the air.

We all became good friends and I must say we had a good time developing the programs. On one occasion, Don Meier was called away from the conference, and while he was gone, someone suggested we play a trick on him. We had a silent dog whistle, with a pitch too high to be heard by human ears, although it is audible to dogs. In the show we were going to demonstrate this phenomenon with an oscillograph, which would show the sound waves made by the whistle, even though no noise was heard. We adjusted the whistle to make an audible sound, and when Don returned, the discussion resumed, and I blew the whistle. Somebody said, "Gee, that's great, I didn't hear a thing!" Somebody else commented, "I didn't either!" And so on all around the conference table. We all agreed we hadn't heard a thing. But Don said, "Come on, you guys; you must be kidding me. I heard it very plainly!" We all looked startled and surprised, and he said, "My Lord, do you suppose I have supersonic hearing?" Well, that was too much. We all broke out laughing. But we have never let Don forget that he has supersonic hearing.

On these live TV programs, before the days of taping, I used to talk not only to the people watching at home but to the animals themselves. I'd ask an animal a question or I'd show him what I wanted him to do or indicate by a movement of my hand or my head how I wanted him to act, and this made a more or less personal relationship that the viewers seemed to appreciate. It was also my natural procedure when working with animals. Even though I feel sure that most animals don't understand what you say, they do understand the tone of your voice and they know you're going to do something that may interest them or involve them.

Working with these animals as much as we did every week, we got to know them pretty well. Many times I could just about tell in advance what an animal was going to do. By anticipating in this way, I could sometimes ask him to carry out an action and have him promptly oblige. An example of this occurred during a demonstration of how opossums climb trees. I knew that when the opossum got to the fork of the prop tree, he would turn and look back over his shoulder. So before he got to the fork, I asked him to turn and he did. On another occasion, when I had a cobra on a table and he had calmed down a bit, I felt that if I moved slightly, he would again raise up and extend his hood, so I asked him to do that before I made my move. I said, "Okay now, let's see that beautiful hood of yours again." And up came part of his body and out went the ribs against the skin to make the characteristic hood of the cobra.

I knew these snakes very well indeed and I knew their limitations. I could work with a cobra on a tabletop with relative safety because I was fast enough to get my hand out of the way even if it was within his striking range. I'd done this over and over again many times, and I knew my reflexes for retreat were quicker than the cobra's strike. Rattlesnakes, the other vipers, and mambas strike more quickly. But an Indian or Egyptian cobra and several other types of cobras stand their ground, elevate the foreparts of their bodies, expand their hoods, and then reach forward in a strike. If they miss, they resume the on-guard position ready to try again.

On Sunday, April Fool's Day 1951, around three o'clock, the show participants were assembled in my office in the primate house at the Lincoln Park Zoo. Don Meier and Jim Hurlbut were there, and my telephone was ringing constantly. My secretary did not work on Sunday, so I had been fielding calls from those practical jokers who on April Fool's Day would give people the zoo's number and say, "Please call Mr. Lion or Mr. Fox or Mr. Wolf or Mr. Al E. Gator," or any other combinations that sounded like a human's name. Jim Hurlbut got in on the act and was having great fun giving flip answers to the callers. This delayed us a little and we didn't have as much time as usual for our rehearsal. When we got to the small-mammal house, where the program was to originate, we started to run through the rehearsal. I was hurrying because I knew we had to finish the rehearsal before we went on the air. I came to the timber rattlesnake which was to have venom extracted from it there on the tabletop. I forgot that I did not have to produce the snake for rehearsal. I needed only to show the camera technicians where the snake would be so they could set their lenses for the shot. But I went ahead with the whole procedure—caught the snake and started to show the position for the venom extraction. I was in such a hurry that I failed to get a good hold on the rattler, and he turned in my grasp and sank a fang in my left middle finger. I slid the snake off onto the floor and one of the keepers put it back in its cage while I reached in my pocket for my knife.

I opened the fang puncture with my knife blade and started to suck out the venom. Gates Priest, photographer for the Chicago Park District, recorded this operation and the additional incisions and suctions, which were applied with suction cups that we always had on hand for this kind of emergency. The bite occurred twenty minutes before airtime. I obviously was not going to do that program and was taken to the hospital before the show started. Lear Grimmer had to take my place. He had been present at the rehearsal, of course, but he hadn't gone through the action himself and had to improvise.

The show was to open with the title of the program on a large card tied to the bars of Judy's, the elephant's, cage. The camera slowly panned from the elephant to the show title, but the card wasn't there. The elephant had eaten it! That was typical of how that program went. In the meantime, I went to the hospital and remained there for three weeks recovering from the effects of the timber rattlesnake bite.

I used to get scratched occasionally, bitten by nonpoisonous snakes, slapped by the tail of a monitor lizard, bitten by baby lions or tigers, but the rattlesnake bite was the worst accident that happened during the life of "Zooparade."

An interesting after-reaction to this episode is the fact that even today I meet people who in all seriousness tell me that they sat there in front of their television receivers and watched that rattlesnake sink his fangs into my finger. At first, I used to correct them and explain that I wasn't on the show that day, that the bite had occurred before we were on the air. But these people are so sure in their own minds that they have seen the thing happen that I now just let it pass and don't try to correct them. Perhaps this shows the power of suggestion. On that show, when Lear Grimmer did the venom extraction, he mentioned that I had been bitten by the very same snake while doing venom extraction, which he then proceeded to do.

The only good thing about the bite of a poisonous snake is that once you recover, you recover completely, except for the scar tissue near the point of injection. I was soon back as host of "Zooparade."

As "Zooparade" continued from one year to the next, and we all became more experienced, Don Meier and I kept trying to improve the quality of the show. We tried to provide better scripts and to have them prepared early enough so that we could look them over at least a day before the show went on the air. This helped greatly and implanted in my memory the right kinds of things to say on the program, but the show still was subject to unexpected goofs when I got my tongue tangled up or an animal refused to cooperate.

A notable goof occurred in a show on animal locomotion—the way different kinds of animals move. A part of this show dealt with snake locomotion. We demonstrated one method, caterpillar traction, with a Gaboon viper that moved largely by this means. The snake really just walked on the tips of its ribs, which run the full length of its body, on caterpillarlike waves of movement of the ventral plates and the skin of the sides of his body. Another type of snake motion, lateral undulation, was shown by a racer. This snake moves from side to side in a series of loops, each loop pushing against an object such as a rock or a piece of grass or a pile of sand. The snake then straightens that section of his body as he changes the position of his loop, forcing his body ahead in three or four long graceful sinuous movements. A third method, sidewinding, was demonstrated by a sidewinder rattlesnake, placed, for our purposes, in sand on a table. Against the shifting sands, the sidewinder lifts up a section of the forepart of his body and begins a roll that continues down its entire length. By the time the roll gets to the lower one-fourth, the snake has enough of the tail section of his body on the ground to lift the forepart off the ground again and start a second roll. This rolling is repeated, and the track of the sidewinder is the most distinctive and unusual of any snake's. It is a series of lateral lines with a tiny hook at the head end. Several other snakes sidewind, particularly those vipers living in sandy desert areas. Nonpoisonous water snakes also sidewind if they're on a slippery concrete floor and can't get traction any other way. But the sidewinder rattlesnake is the best performer I know at this type of locomotion.

A final demonstration of snake locomotion employed the red rat snake that spends part of its time in trees. To show how a snake can crawl out onto a swaying limb or a vine and not fall off, I planned to use a piece of clothesline. Jim held one end of the line and I held the other. I introduced the snake onto the rope, and as it crawled along, I pointed out that with about every five inches of its body, it crossed a loop over the top of the rope, and that gave it traction to caterpillar its body forward. Those body sections that were not supported hung down below the side of the rope, and a little below it, and that lowering the snake's center of gravity so that it balanced. Then to show how really secure this snake was, I said, "Jim, I've done this demonstration many times and it's possible to swing the rope a little and the snake still won't fall off." Jim was always a little nervous around snakes, and when he started to swing the rope, his hand jerked. With that, the snake fell off and plopped on the table! I was aghast. But I laughed a bit and got out of it the best way I could and quickly went on to the next subject. Live TV was bound to have such goofs, and these are the funny things people remember about the early shows.

We had shows about animal superstitions and showed that a hoop snake cannot support itself rolling down a hill and that it is not poisonous, that a rattlesnake will crawl across a horsehair rope, that elephants are not afraid of mice, and that an owl does not turn its head in a complete circle to watch someone walking around it. The old story is that if you find an owl sitting on a pole he will watch you by rotating his head as you walk around and around and eventually his head will twist off. If you walk around an owl so he has to turn his head to the left, he will continue to turn his head to watch you because he can't turn his eyes much in their sockets. When you are at the owl's back and his neck has twisted as far as it can, then quick as a flash (you won't see it if you blink) he will turn his head nearly 360 degrees to the right and watch you as you complete your circle. We also did shows about protective coloration in owls and many other remarkable and little understood aspects of the world of wild creatures.

One Sunday afternoon, as we were beginning rehearsal for "Zooparade," Don Meier told us that at the end of the show we would receive our first award, from *TV Forecast* magazine, which later became *TV Guide*. A few minutes before the close of the program, Jim Hurlbut introduced a young man who read a short statement of commendation for "Zooparade" and handed me a brass cast of a cameraman standing behind a TV camera. A plaque on the wooden base read: *TV Forecast* Annual Award, 1951—Zooparade. It was hard to describe our feelings as Jim and I thanked him for this unexpected honor, but they were feelings of elation and gratitude. My last statement on that day's show was, "Boy, wait until I show this to Bushman and Heinie II." The trophy still has a place of honor on my bookcase.

The next year *TV Forecast* was back with a second award, and soon other awards were presented for Best Children's Show, Best Educational Show, and Best Family Viewing Show. More and more followed until the top of my bookcase and the wall behind my desk were filled with statuettes, plaques, and citations for which I am deeply grateful.

14

The Herds of Africa

"Zooparade" seemed to get better each year. Don and I, casting about for new ways to improve it, hit on the idea of going into the field and photographing nature in the wild. Don suggested this to NBC. He proposed a trip to Africa, where a film crew would take sync-sound 16-mm movies on location, and we would build shows as we moved from place to place in East Africa, including animal dealers' catching compounds near Arusha, Tanganyika (now Tanzania), and eventually on to South Africa. NBC approved.

Before we could leave, however, we had to build a backlog of programs on film that could be played during our absence. So, with a film crew, "Zooparade" went on location in the United States in such places as Jackson Hole in Wyoming and Marineland and Silver Springs in Florida. These trips also served as shakedowns for the equipment and personnel later to go to Africa.

Arrangements were made to fly on SAS. Our route would take us from Chicago to New York, New York to Copenhagen, Copenhagen to Zurich, Zurich to Rome, and Rome to Nairobi. The same plane would continue to Johannesburg. Our return route would be Johannesburg, Rome, Zurich, and Copenhagen. We could thus pick up from my colleagues in those cities the European and African zoo animals I hoped to exchange for the North American specimens we were taking along.

When we arrived in New York to stay overnight, we learned that the shipment of animals, which had come separately by air freight, was being detained in Pittsburgh and that a veterinarian had been summoned because one of the bears seemed listless and not well. I was finally able to convince the authorities in Pittsburgh that the listlessness resulted from tranquilizers, not illness. I reassured the Pittsburgh vet that if he shipped the animals to New York right away, the drowsiness would wear off during the flight. The animals did arrive in time to make the flight to Copenhagen, where they rested in the zoo for two days. During that time we had a charming visit with our colleagues and saw the zoo and its animals. Mr. Boja Benson, the president of the Zoological Society of Copenhagen, entertained us at dinner at a famous restaurant in the Tivoli Gardens. We left one bear, along with descented skunks, American opossums, and assorted reptiles as gifts to the Copenhagen Zoo.

Then we set off with our other animals on our long journey to Africa. First stop, Zurich. And there, the first thing we saw when the plane landed was my colleague, Dr. Heinie Hediger, leading a camel. This curious vision came about because on a previous trip to Zurich I had been impressed by its zoo's barrier-free camel enclosure. There was no fencing and no moat between the camels and their viewers. A four-by-four timber about four feet off the ground served as a railing to keep the camels from walking out of the enclosure. The animals would come right up to this barrier, however, and accept food directly from the hands of the visitors. I wrote about this in my syndicated newspaper column and explained that zoos in the United States dared not take such chances because we felt we would be legally vulnerable if someone was bitten. Dr. Hediger's first words to me were, "Marlin, I want you to see why it is possible to keep the camels as we do. They are perfectly tame. Put your hand up near his face and around his lips; you can be sure he won't bite you." I tried this; Dr. Hediger was right; the camel didn't bite. Maybe the liability laws in Switzerland are a little different from those in the United States.

Our second gift bear was left in Zurich, along with some other North American animals. Our plane stopped next in Rome and we were met by Dr. Bronzini, director of the Rome Zoo. We stayed overnight in Rome, to catch up on sleep and ready ourselves for the long flight to Nairobi. The Rome Zoo received a generous supply of reptiles and amphibians; the third bear was destined for Johannesburg.

I can't express with what anticipation I looked forward to the trip to Africa. I had all my life wanted to visit East Africa and South Africa. I had been to West Africa, collecting the gorillas for the Lincoln Park Zoo. But East Africa I had read about for many years, and I could hardly wait to get there. Our unit manager had gone ahead to make special arrangements, and he met us at the Nairobi airport. We stayed at the Norfolk Hotel and met with the personnel of the safari company who were going to escort us to our filming sites. Jeff Lawrence-Brown was in charge of this operation—a colorful and interesting character who became a good friend.

We started filming in Nairobi National Park, where lions, giraffes, zebras, jackals, hyenas, wildebeests, and a variety of other antelopes, including kongonis, abounded. We filmed a sequence with the director of the national parks of Kenya, Merwin Cowie, who showed us examples of the cruel and heinous devices used by poachers to catch animals in the parks.

Ivory and rhinoceros horn brought high prices in China and India, where the ivory was carved for resale and the horn was ground up to be sold in apothecary shops as an aphrodisiac. Belief in the efficacy of this unproved love potion, hardly needed in these densely populated regions, is in large part responsible for the great decrease in rhinoceros of all species. Animal skins, hooves, horns, and bones were also converted into salable items for the fickle tourist trade—useless things for the most part.

Poachers used various trapping methods. Most common was a wire-noose snare, cleverly concealed and vicious when an unsuspecting animal was caught. Another method involved setting poisoned arrows along a game trail, ready to kill any animal that triggered the hidden bow. A third method was the pit trap. A hole eight feet deep, ten feet long, and four feet wide was dug; long, sharpened stakes were set in the bottom, points upward; and the pit was camouflaged with a layer of sticks and earth. An antelope or buffalo never saw the pitfall; the animal fell through the camouflage layer and impaled itself. Elephants were killed, after prolonged agony, by a painful foot trap. This device, made of bamboo, was circular and just a little bigger around than an elephant's foot. Inside was a ring of sharp, fire-hardened wooden spikes so designed that an elephant stepping into the trap pushed the spikes downward. When he took a step and moved his trapped foot, the ring of spikes penetrated the soft skin of his ankle. Poachers then tracked the animal and stayed within sight of it for days until the wounds caused by the spikes became infected and the elephant became ill. When he finally died the ivory tusks were hacked off with an ax and sold on the black market.

In the Nairobi National Park, we filmed from an open Land Rover. Jim Hurlbut rode in the front alongside Jeff Lawrence-Brown, the driver; I stood behind, holding onto the empty gun rack, along with Chandu, an African game scout who worked for Jeff. On one occasion, we met a pair of lions on their honeymoon. They were lying on a little knoll. Don Meier, in the photography car, took up a position and asked us to drive between the lions and the camera car, stop when we were in the right position, hold there awhile watching the lions, and then move out. This maneuver worked well, but when we stopped to watch, the lion arose and indicated his displeasure at our interruption. I said, "Jeff, if that lion charges, don't start up too fast; otherwise, we'll pull out of the range of the camera." Without even turning his head, Jeff answered, "If that old boy decides to come for us, he'll be on top of us in five jumps!" That made me think about what we were going to do! We carried no weapon in the gun rack, so I looked around and said to myself, "Well, the first thing to go will be Jim's big broad-brimmed hat. I'll sail that at the lion and hope he'll stop and grab it. If that doesn't work, I'll pick up the heavy jack on the floor of the Land Rover, and when the lion gets very close and starts to charge directly at us, I'll throw the jack right down his throat." But, of course, none of that happened. The lion didn't charge, and we pulled out, leaving the two lions in peace.

From Nairobi, we drove in convoy to Amboseli, where Jeff set up a tent camp. We filmed wonderful footage of animals in front of Mount Kilimanjaro, along with the Masai and their cattle and the deep ruts they made crossing the old dry lake bed. The Masai herdsmen here later became an important part of our story.

What a thrill it was for me, who knew zoo animals so well, to see the same animals in nature. Elephants were at Amboseli in some numbers, feeding on the reeds in the swamp and pulling down limbs from the acacia trees. Black rhinos were well represented here; one female named Agnes had grown an unusually long horn and it was she whose male calves were always born without ears. She and another long-horned female of Amboseli were well-known animals. It is mystifying to see a rhino calmly munching an acacia limb studded with thorns.

When we left Amboseli, we stopped for lunch at Namanga, a lovely flower-bedecked, tucked-away rest stop on the main road. A short way up the road lay the border between Kenya and Tanganyika (Tanzania). The gravel road descended into a dishlike plain surrounded by mountains, and before long we passed Longata and Mount Meru, an ancient volcano that had blown its top, leaving some jagged pieces still sticking up. Around Mount Meru and before Arusha, the country developed into lush farmland. One coffee plantation we passed before arriving at Arusha was called Mzuri. We learned later that this plantation was owned by a man from Saint Joseph, Missouri. *Mzuri* is Swahili for "good, fine, beautiful." Another huge plantation grew not only coffee, but also red beans, which were sold in Europe mostly as seed beans, providing a good revenue for Tanganyika.

Arusha was the kind of African town you would expect. As you turned the corner past the native market, you saw cattle and goats and sheep being driven alongside the road. The area swarmed with Africans in their colorful dress. The little stalls and shops sold fresh fruits and vegetables—oranges, bananas, mangoes, gourds, squash, pumpkins, corn, stalks of sugarcane—as well as such articles as old bottles, tin cans, and other utensils. Farther on down the gravel street were low buildings with shops operated by Asians, Africans, and British. The New Arusha Hotel was situated on one side of a circle at the end of the road. It was a low, flat structure with an open courtyard, lovely flowers, and a swimming pool. Hemingway had stayed there, as had Humphrey Bogart and Katharine Hepburn. Near the front of the hotel was a large sign saying: This Point Is Halfway Between Cairo and Capetown. Of course, we had our pictures taken there, as every tourist does. In Arusha Tom Arend, our assistant director, became ill with a mild form of sleeping sickness and had to be hospitalized.

Our next film location was a large ranch operated by Willie De Beer, a South African who had acquired the property after World War I when the original owner, Christoff Schultz, a German animal collector, was arrested and interned by the British. When Schultz was eventually released, his farm was offered for sale and De Beer bought it. Christolf Schultz went to South-West Africa where he continued collecting animals for zoos all over the world. Willie, too, collected wild animals for export to zoos everywhere. He was affiliated with the famous Ruhe family of Hanover, Germany. Willie helped us organize several shows about how wild animals are captured.

One half-hour segment told the story of Willie hunting a giraffe in his 1939 Ford truck. His nephew, Wilhelm, was strapped in a seat on the left front fender holding a long catching pole attached to a lariat, one end of which was tied to the back of the truck. We took out after a giraffe of the size Willie wished to capture and overtook it. Using the long pole, Wilhelm placed the noose over the animal's head. I was riding in the left-hand seat of the right-hand-drive truck, bouncing up and down with no cushions. The door was tied shut with a piece of rawhide, and there were cracks in the windshield. With my hand-held Bell & Howell camera, I filmed the running of the giraffe and the final placing of the rope over its head. Jim Hurlbut was likewise filming with a hand-held camera while standing in the back of the truck bed. Another camera was in a Land Rover racing alongside. Don Meier and his photographer had set up a tripod at the point where Willie expected to catch the giraffe. But the giraffe veered too far to the right and that camera never even rolled. After we had caught the giraffe, Don Meier came up saying, "My Gosh, we'll have to do it again; we didn't film anything!" I assured him that Jim and I and the man in the Land Rover had taken a good sequence of the catch. The sequence finished with the capture of the young giraffe by hand. It showed the hunters taking the rope off the animal's neck and then lifting him, with the combined efforts of six or seven men, into an open crate, especially designed for a giraffe, on the flatbed of another big truck. After he was safely in, a sliding door went down behind the giraffe and the top of the crate was left open so he could stick his neck out. Willie placed a cloth over the animal's eyes until he settled down. He was then taken to Willie's compound, where he was released into a holding yard with other giraffes that had previously been caught. He walked right up to the feeding trough and started to eat.

Willie had built nice holding pens for various animals, with stockade-type wooden fences in front and stalls in the back, where the animals could be locked up at night. One pen contained a group of zebras; another, animals that were a cross between zebra and horse; these hybrids were, of course, infertile. They were sold mostly to

German circuses. They had the brown coat of the horse parent and some zebra stripes. In another pen was a wildebeest; in another, a waterbuck. Then came two white rhinos from the Sudan—young ones which were actually tame enough to ride. An ostrich and several species of antelopes completed the tally. We filmed all the animals, with appropriate comments by Willie, and Jim Hurlbut signed off that program with *kwaheri,* the Swahili word for "good-bye."

Later, after returning home, I told George Vierheller about the two Sudanese white rhinos—a rare species; he purchased them for the Saint Louis Zoo. They were placed in the new large-animal house, which had a spacious outdoor moated yard, something Lincoln Park lacked.

Our camp was set up in a grove of trees on the mountainside, above the animal compound. One morning we noticed a group of people on the road, and learned they were the military. During the night just a mile from our camp, the Mau Mau had killed a black man. The incident brought home to us the determined push for independence in these three East African British protectorates.

That campsite was a lovely one. I never tired of the view of the plain and the rounded knobby hills dotted here and there. The landscape was nearly bare of trees, and the grass was cropped closely by the cattle and goats that grazed there. Our view was to the west, with glorious sunsets.

My seventeen-year-old daughter, Suzanne, with Lorie Meier, Don's wife, toured Europe during most of the time we were in East Africa. But they joined us before we left, saw a little of East Africa, and shared the rest of our tour of the southern part of the continent.

In addition to the staff to produce the program, NBC made it possible for John Crosby, at that time a TV columnist in New York, to accompany us and write stories for his columns. Arthur Shea, a still photographer from Chicago, was along also. He had introduced me to the single-lens reflex camera, and I had acquired one with a 300-mm telephoto lens. It opened up a whole new field of photography for me.

In South Africa we filmed several shows, one at Port Elizabeth at the famous snake park started by Fitzsimmons and another at a bird sanctuary on Bird Island off the coast north of there. Another show dealt with ostriches and was produced at the ostrich farms at Outshorn near Capetown. Our South African filming was finished at the famous Kruger National Park.

During the time we were away, NBC was negotiating with the Chicago Park District about a continuation of their contract. The two sides had come to an impasse, and while still in Africa I was instructed (in a letter from the superintendent of the park district) not to appear in any films or TV shows for NBC. This came as a shock to me. I replied that, as the superintendent knew, I had a contract with NBC and was at that moment in Africa filming ten half-hour TV programs for them. The park district left me no alternative but to resign as director of the Lincoln Park Zoo. This was a serious moment in my life. I had always been a zoo man and was distressed to be caught between the Chicago Park District, which operated the Lincoln Park Zoo, and a successful television program. I discussed this plan of action with producer-director Don Meier. As a result we decided to travel for "Zooparade" to some other zoos before going back to Chicago.

On the return flight, we picked up some South African animals from the Johannesburg Zoo and some European specimens from the zoos in Rome, Zurich, and Copenhagen. With the animals aboard, we flew the polar route on SAS and landed at Los Angeles. Plans had been made for a live "Zooparade" program right on the airport strip when we landed. We had the animals, both Jim and I were there, and so was Don Meier. Our plane happened to arrive at the right hour, and to introduce the program was a Los Angeles announcer by the name of Chet Huntley, a tall, handsome man who greeted Jim and me graciously as we disembarked—me with a monkey on my shoulder and Jim holding a meerkat. For the show we opened other boxes, and talked about the animals that had arrived in them and the many places we'd visited in Africa. The animals were then shipped to the Lincoln Park Zoo, while the "Zooparade" crew went on to Seattle, Washington, to do a live show there from the Woodlawn Park Zoo. The next week we were at the Cheyenne Mountain Park Zoo in Colorado Springs. And there we learned that contractual arrangements between NBC and the Chicago Park District had been worked out. So we came home, and I learned that my resignation had been rejected.

15

To the Amazon: Indians and Pink Dolphins

In 1955, at the end of a five-year contract, Ken-L Ration withdrew, and Mutual of Omaha picked up the sponsorship of "Zooparade." The episodes filmed in Africa had been so successful that we suggested to NBC another on-location filming expedition during the summer of 1957. NBC approved, and the Chicago Park District was willing, so we planned an expedition to the upper Amazon River, with Leticia, Colombia, as our headquarters. Ross Allen, of the Reptile Institute at Silver Springs, Florida, went ahead of us to establish camp and check out the territory. Ross was an old friend of mine. We were both interested in reptiles, and I'd been to Florida on collecting trips many times. Ross set up the arrangements, and a filming crew, with Don Meier in charge, came with us to film several interesting episodes.

Leticia is an outpost town in the low-lying southeastern corner of Colombia, close to the Brazilian border. For one of our episodes we photographed the bizarre hair-pulling ceremony of the Tecuna Indians, who lived in Brazil just a few miles from our camp. These people, sixty or seventy in all, dwelt in huts. Each hut was comprised of a platform about four feet off the ground, with a roof thatched from palm leaves and supported on poles. The sides were open. The Tecunas slept in hammocks swung between the roof poles. In the center of the group of primitive dwellings was a cleared, open space of hard-packed earth, and there the ceremony was held.

The ritual took place when a young girl was becoming an adult. Some days before the event she was locked in a small building, rather like a cage; no one was allowed to see or talk to her except her mother, who fed and cared for her.

On the day appointed for the ceremony there was great activity in the village. We arrived early to set up our cameras. The Indians had made banana wine, and everyone had partaken of it freely. The ritual began when a band of men dressed like demons in costumes fashioned from jungle cloth, came running and screeching into the cleared space between the huts. They were followed by other Indians, some dressed in costumes, others nearly concealed behind large round frames covered with decorated jungle cloth, which they constantly rotated, meanwhile uttering weird guttural sounds. To the music of drums and of a piercing sort of whistle, the yelling crowd paraded back and forth, up and down, across the central plot of bare earth.

Finally the climactic moment came. Three women led the girl from one end of the village to the center of the open space. They placed on the ground a mat woven from palm leaves, on which the girl knelt. The women knelt around her. One of them grasped a small tuft of hair at the front of her head. With a quick jerk she pulled it out. The second woman pulled out another tuft from the side of the head nearest her; the third did the same from the other side. This went on for about an hour. The women jerked tufts of hair from the young woman's head until she was completely bald. Although the process must have been very painful, the girl uttered not one sound of complaint. These Indians are stoic people.

The girl and the women now rose to their feet. The women placed a feathered headdress on her bald scalp and escorted her to a stream at the bottom of a slope. Slowly she entered the water until it was up to her shoulders. Then she bowed her head, placed her face under water, and straightened up again. This ended the ceremony. The girl, though only in her early teens, was now considered to be ready for marriage.

We saw the same young girl three or four weeks later, and I was glad to note that her hair was already beginning to grow back.

We caught a number of animals on this trip, including several fresh-water dolphins, called *bufeo*. These dolphins are white or pinkish in color. In the muddy waters where they live, vision is not important to them, so their eyes are very small; other sensory organs are acute, however, enabling the dolphins to find the fish that are their normal food. We caught some animals in nets and others with a harpoon; in the latter case, the weapon barely penetrated the skin and caused only a small wound that healed quickly. After we had caught a dolphin, we had to get into the water ourselves and lift the animal into our boat. A dolphin could be carried in the canoe for only a short time and had to be rolled on its side so that its own weight would not suffocate it. From the canoe, each animal was put into a pen that had been marked off in the water nearby. These dolphins were eventually sent to Silver Springs, Florida, and were on exhibit in that beautiful tourist spot for some time.

Many of the reptiles we captured were species I'd never had the opportunity to work with before—a bushmaster, a fer-de-lance, and a mussurana. The mussurana is a snake that feeds on other snakes. It's a sizable creature, six or seven feet long when full grown, and black. We were lucky enough one day to see from the top of a small rise a mussurana obviously hunting around a fallen log. Warren Garst quickly set up his camera and filmed the snake as it hunted along the log. As the mussurana got to the other end of the log, its hunting interest obviously increased. Warren was filming when, at the end of the log, the mussurana encountered a fer-de-lance, which it grabbed at the neck with its jaws, quickly wrapping coil after coil of its body around the fer-de-lance, and squeezing it hard. The fer-de-lance was trying to bite, but the mussurana had grabbed the fer-de-lance so close to its head that it could not bite. (It is my understanding that the mussurana is immune to the venom of the poisonous snakes it eats, so the bite would probably have made no difference.) When the fer-de-lance became limp, the mussurana relaxed its coils, straightened out, got hold of the fer-de-lance's head with its jaws, and swallowed the victim head-first. A wonderful sequence for our film. The fer-de-lance was a little smaller than its adversary—about four and a half to five feet long. It filled the mussurana's body cavity.

With Ross Allen, who was an accomplished snake catcher, we also found a bushmaster, the biggest poisonous snake of tropical America. This specimen was about ten feet long. With our snake hooks we were able to bring it out from the thicket into an open area, and then maneuver it back and forth as it struck at us from various angles. Avoiding its long lunges we finally caught it and put it in a flour sack, our usual field container for a snake. Quite a bit of activity was involved in getting the bushmaster into the sack because the snake was so long and struck repeatedly. One man would pull the snake back from the man it had just struck at, and then the snake would turn and strike toward him. So it went, back and forth, several times before we were able to pin the snake's head, grab it by the neck, and drop it in the bag.

It was fascinating to see how the Tecunas and other nearby tribes lived. Their dugout canoes, in Spanish called *cayugas*, are so shaped that they have an angle about a third of the way back from the bow. This is to allow paddling from the bow, using big, heart-shaped paddles formed from single pieces of wood. The Indians paddle from the bow because during the high-water season, when the whole forest is flooded, they are able from that forward position to part the branches of the trees and pull their way through the forest in a manner not possible from a stern location. This is the strangest way of propelling a canoe I have ever seen.

The Tecuna village was two thousand miles from the mouth of the Amazon, and at that point the river, the biggest in the world, was two miles wide and two hundred feet deep. There was a rise and fall in the river, between the rainy and dry seasons, of about thirty feet. When the rains came, the water overflowed through small channels, ten to fifteen feet wide, which carried the overflow into a lake back in the forest. When this lake filled, similar channels at its other side led the water into another lake, and so on into still more lakes through the forest. These lakes became reservoirs during the rainy season. When the dry season came and the water started to recede, it flowed from the lakes in the other direction and helped to maintain the flow of water in the Amazon. It is variously estimated that a third or a fifth of the fresh river water in the world flows down the Amazon. It is an enormous water system, and I am alarmed by talk of leveling the Amazon forest. In many parts of Brazil forests are already being felled to make farmland or other commercial property. One visionary even wants to make a lake out of the whole Amazon complex. Why man would want to disturb that glorious tropical rain forest and the millions of creatures and plants that live there, escapes me. I hope it never happens and that the destruction will soon cease. There are many things not yet discovered in that vast tropical rain forest, the most extensive in the world.

Another fascinating aspect of this part of the world is that many of the rivers interconnect. You can go by boat from the Orinoco into the Rio Negro or the Amazon, and from there south into other streams that flow still farther southward. Because of this interconnection many of the fish and other aquatic creatures are common to all these rivers. More than four hundred different species of catfish inhabit those waters. One is more feared by the Indians than even the piranha—the fish notorious for its voracious feeding on the flesh of anything available. We caught one of these big catfish on a large shark hook baited with a whole dead chicken. The hook was bent out of shape by the fish, which was six feet long—a comparatively small individual, for some grow somewhat larger. Every Indian dreads these catfish. He believes that if his boat should capsize in the Amazon, where the monsters mostly live, they would attack him in the water, pull him down, and devour him.

The Yagua Indians lived upstream from the Tecuna village. The men of this tribe wore skirts made of leaves. The Yaguas are famous for the blowguns they make out of heavy ironwood. About eight feet long, these blowguns are bound with split jungle vines and can be aimed with amazing accuracy. With a blowgun a Yagua shoots a ten-inch dart made of split bamboo having a fire-hardened tip dipped in curare, a deadly muscle poison. A little wadding of kapok down is wrapped around the butt of the dart. With these darts the Yaguas kill monkeys, birds, and other animals for food. On one occasion we marked a small spot on a log and asked a Yagua hunter to shoot at this target. Standing about fifty feet from it, he raised the blowgun in his hands, sighted down its length, lined it up, blew one quick puff; the dart hit exactly in the center of the spot we had indicated.

We too tried shooting the blowguns and became fairly accurate, but never acquired the expertise of the Yaguas.

After two eventful months on location, we had filmed many reels of fine material for "Zooparade," and prepared to depart from Leticia. Before leaving, I was able to buy a blowgun, darts, a quiver, and some kapok in a little pouch such as the Indians use. The darts had curare on them, so I had to be careful with them. I brought this arsenal from the upper Amazon home with me, and still display the components in my living room in Saint Louis. They make a fine conversation piece. And sometimes I put a dart—from which I have carefully removed the curare—into the blowgun and show off by firing it right through a *Time* magazine from a distance of about twenty feet.

16

In Quest of the Abominable Snowman

We returned home from the Amazon to learn that "Zooparade" had been canceled. Don Meier continued his staff job as producer for other programs at NBC-Chicago, and I continued as director of the Lincoln Park Zoo. Don and I had become fast friends and saw each other frequently. We talked about starting another television program, knowing full well that when a program like "Zooparade" went off the air it was virtually impossible to bring it back. Videotape had been developed, and TV was changing; shows were rarely done live. Between us, but largely through Don's expertise in television, public speaking, and story-line development, Don Meier and I conceived the format for "Wild Kingdom." Don wrote up a description of the program, formed a company called Don Meier Productions, resigned from NBC, and found enough money to develop a pilot film.

A young man we had met at John Hamlet's bird-of-prey exhibit in Florida came to the zoo one day to tell me about his recent expedition in South Africa, during which he had collected a group of falcons, eagles, and other raptors. All his birds were broken to jesses and he was training them for a lecture tour. This young man was named Jim Fowler, and his raptors seemed like a good subject for a pilot film. We filmed Jim with his Lanner falcon and his Marshall eagle to show how a falconer trains his birds.

During the summer of 1959, I had a call from Don Meier that went something like this: "Hello, Marlin. What do you know about the Abominable Snowman?" Being an honest zoologist, I said, "Not really very much."

He said, "That's a pity. I've just spent the whole day with Sir Edmund Hillary. He is going to the Himalayas at the head of an expedition to investigate the Abominable Snowman, and he's looking for a zoologist."

I cut in with, "Hey! Wait a minute, Don. Under those circumstances, I have a strong interest in the Abominable Snowman!"

I had read much about the Snowman, or yeti, and was indeed interested. I met Sir Edmund Hillary, of Mount Everest fame, and we talked about the Snowman and methods to be used to try to unravel this fascinating mystery. The expedition, which would also investigate questions concerning the acclimatization of mountaineers at high altitudes, was being financed by the *World Book Encyclopedia*, published by Field Enterprises. The expedition would take three or four months and would be a rugged walk-in journey. Some mountain climbing would be necessary, as the route led over high passes. There would be another zoologist, Dr. Larry Swan of San Francisco State College, who had spent some time in India as a youth and was a high-altitude ecologist. I mentioned to Hillary that I was fifty-five years old, though in good shape. Did he really think I would be physically able to do this? His reply was, "I've been watching you walk down the street with me; I think you'll have no trouble at all."

I had long ago developed a technique of walking as fast as I could for exercise, and I guess that was what Ed Hillary had seen when we walked down the street together. Before he left Chicago, Sir Edmund told me he had definitely chosen me to head that section of the expedition that would investigate the Abominable Snowman.

Despite Hillary's confidence, I decided to put myself in the hands of a good physical education instructor and start working out. The head of the recreation department of the Chicago Park District put me in touch with Al Benedict, the physical instructor for Lake Shore Park. Al suggested late afternoons as a convenient time for both of us. I bought gym shoes and gym clothes and started with regular setting-up exercises in the field house. Then Al took me outside to the quarter-mile track and said, "We'll jog down the straightaway and walk around the curve at the other end and then trot down the other side and walk around that end. We'll make just one round today." That's how my running started. At first, I couldn't have run a mile, but before my training was finished, I had no trouble running a mile at a pretty good clip. Al started me on weight lifting, rings, and horses and bars. We swam often both in an indoor pool and in the lake just off the park. Al put me on waterskis and on a disk (which is even harder to control) and on a trampoline. I had nine months of this kind of physical training, and I got in really good condition. To keep my legs strong, I rode a bongo board several times a day.

During that year, I had many communications from Ed Hillary, outlining plans and discussing possible locations in which to search for the Snowman. At last, a plan and a map of the route of our expedition were worked out. With a photographer friend, I developed a thread release for a Rolleiflex camera that would set off the camera and flash at the same time. We tried the invention out in the zoo at night; strung about six inches from the ground, the thread was easily tripped when an animal walked across it. I bought twelve cameras and twelve small tripods to set them on, field darkroom equipment, and a changing bag so we could develop the film and make prints in the field. I bought a capture gun and some smooth wire that could be used in making crates to hold an animal, just in case we saw one and were lucky enough to anesthetize it. Twelve viewing scopes and twelve pairs of binoculars were added to my store. I took along a camera case with single-lens reflex 35-mm camera, a 150-mm Kilfit telephoto lens, a 300-mm telephoto lens, and a 50-mm macrolens also made by Kilfit. I had sixty-five rolls of Kodachrome film, each with thirty-six exposures. I took a 16-mm Bell & Howell movie camera with four lenses, a tripod, and a stock of movie film. I would, in addition, be operating one of the expedition's movie cameras, also a Bell & Howell, for a planned film of the expedition to be produced by Fred Niles Communications Center in Chicago.

The year 1960 was a most important one for me. Eight months of that year I spent preparing for the expedition, and most of the balance of the year I was on the expedition. But the most important event of my whole life also happened in 1960. My first marriage to Elise had fallen apart, and in 1953 we were divorced. In 1960, after I had been alone for some years, I met again an old friend from our years in

Buffalo. Carol Morse Cotsworth was divorced and had three children—Alice, Fred, and Marguerite. Carol and I were married on August 13, 1960, and she and the children became the family I had always wanted.

Two weeks later I left to join Ed Hillary in Kathmandu, capital of Nepal, the country of the Himalayas. Carol says I spent our honeymoon with Ed Hillary, but I've made it up to her many times since, I hope. In 1961 she accompanied me when I had to go back to the Himalayas at the end of the expedition. From there we went on around the world to Ceylon, Thailand, Singapore, Hong Kong, Japan, Hawaii, and back home to Chicago.

My plane for Kathmandu took me first to London, where I had to stop for several days for a complete physical examination. When I walked into that laboratory, there was a man on a bicycle with a tube in his mouth attached to a balloon-like object; he was pedaling like mad. When he was finished, it was my turn; and the next hour took me through a series of tests. They were vigorous, but I passed. I was glad the high-altitude physiologist who examined us didn't see me walk down the stairs on my way out. My legs gave way and I had to hang onto the handrail.

In London I was introduced to Barry Bishop, mountaineer-photographer for the National Geographic Society, who was to accompany us on the expedition. He and I flew together to New Delhi and then on to Kathmandu. We joined Ed Hillary at the Royal Hotel, where we met a good many of the members of the expedition. This hotel, in the front of an old Rana palace, was operated by Boris Lissanevitch, a charming man who had had an adventurous career and ran a good hotel with a fine kitchen. When the king of Nepal entertained, it was Boris who prepared the food. In a little room on the second floor up a great marble staircase was a bar called the Yak and Yeti. Its open fire in the center of the room was welcome on cold evenings.

The boxes of equipment were beginning to arrive and were placed around the edge of the hotel's tennis court. There they were opened and checked. Then each person had to portion out his own gear into sixty-five-pound loads. That was the limit set by the union, the Himalayan Society, which supplied the porters to carry the equipment to our base camp. We were issued mountaineering equipment as well—boots, jackets, balaclavas, ice axes, sleeping bags, inflatable mattresses, backpacks, shirts, trousers, crampons.

The job of repacking my equipment was completed in a few days. That left me time to sightsee around the Kathmandu area. Never had I seen such a fascinating city. On one of my free days in early September the Festival of the Little Living Goddess was held. The central figure in the event was a young girl, six or seven years old, who had started her training for this great event of her life at the age of two or three. From then on she had lived with nuns, apart from her family, and had been educated for her role as Living Goddess. At last she was ready for her great day. Leaving the nunnery accompanied by special dignitaries, she walked through a dense crowd a short distance to her chariot. She was resplendent in an elaborate costume, and wore a golden, jeweled crown. Her face was heavily painted and her eyes accented with mascara. Her chariot was a great hand-carved, richly painted cart with huge wooden wheels. She was helped to the throne in the chariot, and then twenty or so men lifted the long tongue of the conveyance and pulled it slowly through the enormous crowds. So many people filled the narrow streets that you wondered how the ceremonial vehicle got through. Finally it turned a corner and passed in review before the king and queen, crown prince, and various ministers and dignitaries of the government.

On another sightseeing day I visited Swayambhunath, the sacred-shrine temple where Buddha taught, beautifully situated on top of a hill, and also Pashupatinath on the Bagmati River, site of a Hindu temple to the Golden Bull. I visited Bhatgaon, the ancient Newar capital of the great valley of Kathmandu, and was able to photograph the Golden Palace Door. I saw a child bride of thirteen in her bridal costume being carried in a palanquin to her wedding. In a little village called Bodhnath, I saw the largest stupa in the world—a great white-domed structure with the four faces of Buddha enshrined under a canopy at the very top.

But most of my free time I spent roaming the streets of Kathmandu and taking pictures. From the roof of our hotel, on clear days, I could see high mountains in the distance, with an occasional snow-capped peak looming against the sky.

Eventually the day of our departure arrived. We had our last ride in a vehicle, a broken-down jeep, over very difficult roads to a field on the outskirts of Kathmandu. There the various packages of equipment were arranged in neat rows. Nearby our porters sat on the grass or milled about waiting for the final departure. I found my equipment, thanks to Ang Pemba, the Sherpa assigned to look after me, and before long the porters lifted their packs. With the tump strap in place around the pack and over his forehead, each man adjusted his load to a comfortable position and started off. They formed a line, and Ang Pemba, some of the sahibs, and I fell into line too, and the expedition started along the trail.

We passed through a village and stopped to photograph the fabulous architecture of the early Newars and of the Buddhist temples. We did not travel far that day, but pitched camp alongside the trail, near where an American missionary and his wife were living. Our Sherpas put up our tents, inflated our mattresses, and with the help of some of the sahibs, cooked dinner. I crawled into my half of a small tent and went to sleep thinking, "Well, if I've forgotten anything now, it's too late."

Next morning, we were on our way early. Following a well-defined trail, we trekked past fields of wheat ready to be harvested, past farms and through villages, up and down hills, up and down, up and down until my feet began to feel sore. I saw a brightly colored rock that looked unusual, picked it up, and waited until Larry Swan, the high-altitude ecologist, had caught up with me. I said, "Larry, you studied geology, didn't you? Can you tell me what this stone is?"

Larry gasped, "Later, Marlin, later! I'm just trying to survive!"

About dusk, Larry and I and our two Sherpas stopped to rest. Larry and I had just about had it. Our shoulders were raw from the straps of our packs. My feet were blistered. Seeing my discomfort, Ang Pemba took part of my pack and tied it atop his own. Larry was able to lighten his load a little too, with the help of his Sherpa, and we started off down a winding, rocky trail. It was getting darker all the time, and we had resorted to flashlights before at last we reached a bridge across the Sunkosi River. A short distance upstream the rest of our group was camped: two hundred porters, about a dozen sahibs, and a heavy sprinkling of Sherpas.

Camp was already set up and dinner was being prepared. As soon as my tent was put up, I crawled into it and lay there resting until dinner was ready. I crawled out and managed to get something to eat. When Ed Hillary came around to see how I was doing, I explained about my blistered feet. He gave me some adhesive that was especially designed for blisters.

We were awakened in the morning with tea and before long were on the trail again, strung out for about a mile. We stopped, about ten o'clock, as we had the day before, at a likely place near a stream, where we cooked breakfast. An hour later we were back on the trail, climbing a mountain, traversing a pass, and descending the other side of the mountain to a stream. Some streams we crossed had foot bridges that swung under their supporting cables; others had logs laid across to a midstream rock and then more logs laid from the rock to the far shore. We climbed some steep rock slopes with steps cut into them, and we frequently passed houses with small terraced farming areas around them. We were changing altitude by as much as 10,000 feet per day. This trek was part of the acclimatization program and was also designed to toughen us for the high mountains we would encounter later. It also was the only way we could get to our destination. It took us twelve days to walk to Beding, a small village in the Rolwalling Valley, a distance of about 140 miles. We arrived in Beding after having camped the night before on top of a saddle which we had reached by hiking up a slippery trail in a forest during a rainstorm which turned to snow. The night before that, we had camped on a farm near a river at a low altitude, and in the late afternoon some of the farmers had talked with us and I had inquired about the local animals.

Through an interpreter, I found out that a small red panda, the lesser panda, was native there. I asked the farmers, if they located a panda, to cage it and send it to Boris Lissanevitch at the Royal Hotel to keep until I returned some months later. As it happened they did bring two pandas to our first base camp, set up next to the village of Beding, and I was able to send the animals back to Kathmandu. Those pandas eventually reached the Lincoln Park Zoo and made a fine exhibit there.

We had with us on the expedition a first-class interpreter. Desmond Doig, an Englishman born in India and raised speaking Hindi. He had been in a Gurkha regiment during the war and spoke Nepalese as well. A writer for the *Calcutta Express*, he was both interpreter and official reporter for the expedition. It was he who prepared the press releases that were carried back to Kathmandu by two runners and delivered in about a fourth of the time it took us to make the journey. When we reached Beding, Desmond's empathy with the people and knowledge of their religion and background enabled him to make friends easily and to ask about the yeti. He explained about our interest in learning as much as we could about this creature and announced we were prepared to pay for any bits of evidence—skulls, bones, skins, claws—of the Abominable Snowman. Word was out, and Desmond maintained his contacts. We went to the *gompa*, or local temple, the day of our arrival, and under Desmond's tutelage and timing, sat through a religious ceremony punctuated with horns, bells, and gongs. At the proper moment, Hillary made a contribution from our group and presented a white silk scarf to the head lama.

I searched for areas to set up my self-tripping cameras, and Ed Hillary scouted lookout places for the mountaineers who would man these stations. Each lookout was equipped with binoculars and a viewing scope on a tripod and was asked to scan the valley, the opposite mountainside, and the area nearby for any moving object and then to zero in with the viewing scope to identify it. I placed cameras under ledges, at the entrance to a cave, and on trails through the forested area of the mountainsides; each day I made an inspection trip. The location seemed to me an ideal place for the yeti: At the bottom of the valley was a swift-running stream constantly supplied by the melting snows above; a deciduous forest bordered the stream on both sides. About 500 feet above the river, this deciduous forest gave way to bamboo forest, which extended another 500 feet. Beyond this a rhododendron forest extended about the same distance, and above that lay a grass area and a region of lichen-covered rocks. Beyond that was snow. Food was available on that mountainside, and there was evidence of small animals on the trails.

I had an attachment that allowed a camera to be coupled to a viewing scope so pictures could be taken through the scope. Using one of these scopes a viewer could tell the difference between a dog and a sheep three miles away.

We spent twelve days in Beding and didn't see a yeti. Nor were any of the cameras tripped. But something else did happen. Desmond found out that there was an alleged yeti skin right in the village. It belonged to a lama. At first it was not available for even Desmond to look at, but eventually he was allowed to see it, and then he bargained to buy it. In the end he won. Ed Hillary agreed to the price, and the skin was ours. We spread it on the ground—the skin of a rare Himalayan blue bear, *Ursus arctos pruinosis*. Our Sherpas and some Tibetan refugees in the village assured us it was the skin of the big yeti, the chuti; they were afraid of it. The chuti killed yaks, they told us, and sometimes carried off maidens. Something terrible could happen to you if you even looked at it. Shortly thereafter, we acquired another bear skin, this one from a Himalayan black bear. When we laid the skins side by side and asked the Sherpas what the blue bear skin was, they said, "Oh, that's the chuti, the big yeti—terrible, terrible!"

"And what is this skin over here?"

"That's just a black bear."

Why they didn't differentiate these two as just different kinds of bears, I'll never know. They positively identified the blue bear with the yeti. Later on we got two more blue bear skins, and wherever we showed them to the Sherpas and Tibetan refugees, we heard the same thing: "These are the chuti, the big yeti."

At the end of twelve days, Hillary thought we should move on. We climbed another 2,000 feet to the summer village of Beding, where the inhabitants moved their yaks, goats, and sheep for the summer and where they planted their potatoes. Temporary houses had been built of piled-up stones; the houses were roofless and the villagers used mats to cover the huts in the summer months. The fields were outlined with stone fences—just rocks piled together to form a barrier. When the villagers wanted to put their animals inside, they rolled a few rocks out of the way, herded the animals in, and replaced the rocks. Again we set up cameras and observation posts and assigned mountaineers to scour the area. Again, we found no concrete evidence of a yeti. None of the cameras was tripped and none of our observers saw a yeti.

After two weeks Ed decided we would move further up, this time to a sacred lake at 16,000 feet on the edge of the Rolwalling Glacier. We again set up cameras and observation posts; again our scouts scanned the mountains nearby. One day they excitedly reported yeti tracks in the snow. We all ran out to see. But the tracks were too badly melted to constitute hard evidence, and although we took a plaster cast of the best-looking one, we didn't feel we had found a really valuable clue.

We spent another two weeks at this third location, and then moved up another 2,000 feet to establish a camp on the upper portion of the Rolwalling Glacier not more than two miles from the Menlung

Glacier, the place where Ed Hillary and Eric Shipton had parted company on their reconnoitering trip for the climb up Mount Everest. Ed had said, "Well, I think I'll go on down the Rolwalling Glacier and see what the view is from there." And Shipton had replied, "Okay. I'll go down the Menlung Glacier and I'll see you in Kathmandu."

Not more than a mile beyond where they parted company, Shipton came across some tracks and took perhaps the most famous picture of a yeti footprint. We were in the proper vicinity and perhaps we too would find yeti tracks. The next day, Ed Hillary took a group up to the point where he had left Shipton; Desmond Doig, Tom Nevins, and I, with several Sherpas, climbed 500 feet up a black rock cliff to a saddle. On this saddle, our Sherpas said, was a series of yeti tracks they had found the day before with sahib Larry Swan. Our Sherpas, who had discovered the tracks, pointed them out and assured us, "Yes, yes. These are indeed yeti tracks!" I took pictures; we measured the footprints; Desmond made sketches of them; and we started to follow the trail. It wasn't long before we noticed that no two footprints were the same. Some were just ovoid depressions in the snow; some had little projections at the end of the ovoid; some were just blobs in the snow. We photographed a long series of the prints, with a ruler placed in the picture to show each print was about ten inches long. But again, there was great variation in their size.

It soon became evident that something had caused these tracks to change. They weren't the same as when the animal had made them. We backtracked down a north slope where the sun seldom shone, and there we came upon the tracks of the animal that had made the "yeti" prints. The unmelted tracks were clearly identifiable as the footprints of a fox or a foxlike animal that had walked up that slope and across the saddle. His tracks, where the sun could reach them, had melted or ablated into the shapes we had seen. When surface snow is broken, it is a well-known fact that ablation does occur, and at high altitudes (we were at 18,500 feet), with a strong sun and a receding snow line, the snow vaporizes into the air. We continued to follow the fox tracks in the direction in which the animal was headed, and soon we could see that tracks made in a place where the sun shone down the length of the ovoid depression in the snow grew toelike projections. But at a place where the fox had turned a corner and was going ninety degrees in a different direction, the toelike projections grew out the side of the print rather than at the end. We continued to examine the footprints, and to photograph, measure, and sketch them; we all came to the same conclusion. When Ed Hillary and his group returned from the glacier, we showed them our find. Ed and his mountaineers examined the tracks and agreed that our explanation was plausible and logical.

Next morning at camp, I noticed a raven hopping around above the thrown-out garbage. I reached for my camera, and as the bird hopped away from me, I saw his tracks being made in the snow. Later in the day, I again saw the raven and again tried to photograph him. And then I noticed that the tracks he had made in the morning had already ablated into much larger tracks—not as big as the ones we had seen the day before, but nevertheless ablated so much that they looked like the tracks of some other larger animal. This convinced me that our analysis of the "yeti" tracks was correct. Later, on the way down from this camp, we saw two more series of tracks crossing the glacier; these had the same conformation as the "yeti" tracks we had first found. All showed ablation as they neared the top of the glacier where the sun shone brightest, but in the shadows were clearly identifiable as fox tracks.

We had completed our work in this area, and the time had come to go to Khumjung. On the way, we stopped at Namche Bazar, the chief town of the Sherpa country, where there was a radio station and a small military attachment. We paid our respects to the government official in charge, spent the night there, and the next day climbed a trail into a beautiful area. As we started our descent into a lower portion of a canyon, we turned a corner, and there before our eyes was the most enchanting village I'd ever seen. This was the village of Khumjung, and it was just as I had always imagined Shangri-la should look. We descended the trail to a stone gate, past some Buddha carvings on rocks, into a small open valley, and into the town. There were no streets—just a series of stone houses with some fields, stone fences, and a little valley entirely surrounded by mountains. To the east was Ama Dablam, the fish-tailed mountain, and straight up the gentle slopes of the village was the villagers' own mountain. In their opinion and in their religion, each mountain is occupied by a god, so each one is sacred.

We pitched our tents in a field, down a ways from the village, and then met the local officials. Several of our Sherpas were from this village, and Ed and some of his mountaineers had been here several times before. They knew and were friendly with the Sherpas. We visited the *gompa* (temple) and met the elders. Desmond asked to see the famous yeti scalp, and it was brought, wrapped up in cloth. It was unwrapped and one of the elders, Chumbi, put it on his head. I photographed him. Then Chumbi gave the scalp to Hillary, who put it on his head. We passed the yeti scalp around and all had our pictures taken wearing it.

Then we got down to the serious business of examining the scalp. We looked it over carefully. I photographed it from every angle; I used my macrolens to show the inside structure as well as the bristling hair that ran up from its front across the peaked top and down the back. We measured the scalp and asked the elders questions about it. They told us, through Desmond, that this scalp was 240 years old. They had acquired it at a time when there were lots of yetis on the mountain. The yetis were killing yaks and goats and had become bold enough to come right into the village. The Sherpas decided the yetis would have to be destroyed. A number of Sherpas, including the lama, went up the mountainside with some *chang*, the local beer. They carried with them their swords and some imitation swords made of wood. They reached a place where they saw the yetis watching them from above. There they stopped, opened their *chang* bottles, and drank. Soon they pretended to be drunk, picked up the wooden swords and pretended to fight fiercely. This lasted for some time; then they laid down their real swords and the bottles of *chang*, and taking the wooden swords with them, hid behind some rocks and watched. The yetis came down, drank the *chang*, seized the swords, started fighting, and so killed each other—all but one, and he was terribly drunk. The lama rushed out of hiding, picked up a sword, and killed the surviving yeti. His scalp was cut off and had been kept in the village of Khumjung ever since.

Desmond explained to the elders our interest in any evidence— skins, teeth, claws—of a yeti. That evening and for several days later, we discussed that scalp. We decided it didn't look like a true scalp from an animal. There were physiological reasons for our skepticism. In a gorilla's skull, for instance, the brain portion of the skull was forward of the sagittal crest. But the "yeti" scalp's owner had had an even, high dome leading straight up to the top of its crested head. Those animals that need sagittal crests need them for muscle support, and the scalp showed no evidence of such a need. Also odd was the row of stiff bristles running from the eyebrows over the top of the head and down to the neck. Usually stiff bristles are on an animal's back, and usually the bristles are erectile. We had also noticed some little holes inside the skin that appeared to us to be holes for pegs. We surmised that the scalp had been made by stretching a skin over a wooden mold and pegging it in position until it dried. We thought, also, that the henna color was artificial—it looked phony. We surmised that the hair had been dyed.

Desmond continued to make friends with the Sherpas at Khumjung and to talk to them about yetis and ask for other bits of evidence. These requests produced a couple of claws with some black hair still attached. We identified them as claws of a Himalayan black bear. A "yeti" skin was found to be the skin of a Himalayan serow, a type of goat-antelope. It was a gray-black skin, gray on the sides with some black guardhairs, and blacker toward the top of the back. Running down the length of its back was a row of long, black bristles, just like those on the Khumjung "yeti" scalp. Perhaps this was the answer.

We had heard about another scalp in the village of Pangboche, so Desmond and I and a couple of the others hiked over to that village, which was much closer to Mount Everest. The lamas there showed us the scalp; we found its measurements were identical to those of the Khumjung scalp. It too had little holes, as though it had been pegged to a mold. It too was ancient, but the hair had not been so badly worn off. It too was a phony henna color. The two scalps might have been made on the same mold.

The lamas also showed us what they described as the skeleton hand of a yeti; this turned out to be human, probably a hand of one of the ancient lamas.

We bought two more serow skins and decided to sacrifice one to see if we, too, could make a yeti scalp. Desmond hired a lama to make a wooden mold the same size as the inside of the "yeti" scalp. When that was finished, we soaked a dried serow skin in wet mud until it was soft and pliable. We then stretched it over the mold, tacked it in place, and let it dry for a week. Then we pulled out the tacks, took the skin off the mold, and found we had a pretty good replica of the "yeti" scalps in the villages of Khumjung and Pangboche, except that ours was new and its hair was much thicker than the hair on either of the others, which were old and had been worn in religious ceremonies for many years. Our new scalp was also the wrong color; it was not hennaed. Desmond asked one of our Sherpa friends how we could make this scalp the same color as the one in the *gompa*. "Tomorrow," the Sherpa said, "I bring you something." He brought a little rag containing some dried leaves, put the leaves in boiling water, and soon had a very nice henna-colored dye. We hennaed half of the new scalp just to show it could be done, but we left the other half its natural color to show that this was an artificially dyed scalp—or more properly, a hat. I believe these headdresses had been made as religious objects, to be worn in some religious ceremony. We frequently saw pointed hats in the Sherpa villages, and pictures from Tibet show lamas in peaked, pointed hats.

We had a conference, went over all the work we had accomplished so far, decided that we had the right answers about the origin of the two scalps, but that it was our word against the assertions of all the people who claimed the scalps were yetis'. We determined to try to get permission to take the Khumjung scalp back to the United States and Europe where it could be scientifically examined. Our theory could then be verified by the scientific community.

Desmond and Hillary opened negotiations with the four elders of Khumjung and Khunde. The elders at first objected to allowing a sacred object to leave the village; they feared its removal would loose some terrible catastrophe. We offered to take with us a custodian who would look after the scalp and be responsible for it, and we offered to help raise funds for a much needed new church. We asked them to name a price for allowing us to take the scalp away for two months. They named a figure, but imposed a six-week time limit. Our pleas for extra time left them unmoved.

"No," they said, "we cannot let it be gone more than six weeks. And besides, we have to get the vote of all the people in the two villages for approval."

We were forced to agree to the six-week limit, and they promised to hold the voting that same night. A few of us were invited to eat dinner and to spend the night at a Sherpa house while the vote was being taken. When we arrived at the house, we opened the door and walked into the stalls of the yaks and sheep and goats. Off to the right of one stall was a ladder leading up to an opening on the second floor. This was where the Sherpa family lived. A patch of mud had been laid down on the floor in a four-foot square, about four inches thick, and on this the Sherpas built their fire. There were two windows in the building, close together on the south side. There was no chimney and no other openings except chinks between the hand-made shingles. The place soon filled with smoke. We sat on a bench next to the two windows and were served *rakshi*, a stronger drink than *chang*. It is actually distilled from *chang*, which is made from barley mash.

After dinner, we waited for the results of the voting. But we didn't find out until morning, when the four elders came staggering in. They had visited each house in the village and had tallied every person. When you visit a Sherpa, you sit down and have a drink. There were a lot of houses in the two villages, and the elders had had a lot to drink. The vote was affirmative. We now needed only to increase the agreed-on amount of money for the church. I asked what percentage of the vote was required to reach a decision. They said, "Why, one hundred percent, of course. We're a democracy!"

We were grateful, and happy to know the villagers had appointed Chumbi as the scalp's custodian. Desmond and I had one more investigation to conduct. We had to visit the head lama—the lama Rimpoche—of the monastery at Thyangboche. We wanted to consult him about a report from a previous expedition which claimed to have visited his monastery immediately after a man had been killed by a yeti. According to proper custom, we presented the obligatory scarf to the lama and made a contribution to the monastery. The lama then invited us to tea and also offered us some canned corn and beans left by a previous expedition years before. He had the cans opened and we managed to swallow a little, although they had spoiled. Desmond spoke to the lama in his own language, telling him of our interest in the yeti and in the report of a death. The lama said, "Oh, yes. I remember when they were here. Yes. They had two dogs. But no man has ever been killed here by a yeti. They did photograph the cremation of a man who had died in the village, but he died of natural causes. He wasn't killed by a yeti." Then he said, "I don't know whether there really are yetis. I've never seen one, but I am a young man. Let me call an older lama and question him."

We then met a man who had been at the monastery for fifty-five years. He said the same things. "Yes, I remember when that expedition came through here. They had a couple of dogs with them and they did photograph a cremation here. But that man had not been killed by a yeti; he had died of natural causes." When we asked his views about the yeti, he answered, "I don't know. I've never seen a yeti. We hear from the Sherpas and the people nearby that there are

yetis in these mountains. They say they have heard them. Perhaps—
some say they've seen them. But I have questions in my mind about
yetis."

Some people like to keep legends alive, particularly such a fascinating legend as the yeti. Others have personal reasons for doing so.

The visit to the Thyangboche monastery completed our investi-
gations, except to get verification of our theory about the Khumjung
scalp. Plans were made for our departure and instead of an eighteen-
day walk to Kathmandu, we made the trip in nine days. The day after
our arrival, we made an appointment to show the scalp to King Ma-
hendra of Nepal. He granted us permission to take the relic out of the
country, and the next day we flew to Calcutta. As the plane climbed
from the Kathmandu airport, we overflew Khumjung. Chumbi's face
was glued to the window, looking down particularly at his village's
local mountain. He had never been higher than his own mountain and
was trembling at the thought of how the god must feel about a mor-
tal's being higher than he.

From Calcutta we flew to Chicago, where we arranged to show
our yeti objects to a group of scientists who were to gather at the
Field Museum. At the appointed time, we arrived, bringing the
Khumjung scalp, the fake scalp we had made, the skin the fake had
been cut from, the two other serow skins, the skins of the Himalayan
blue and black bears, and the bear claws. The Field Museum personnel had laid out a representative collection of Himalayan animals, and
at least fifty scientists were on hand to participate. They examined the
skins and scalps we had brought and compared them with the museum's serow skins. Theirs were very similar to ours. Chumbi graciously allowed a small scraping of skin and a few hairs to be removed
from the Khumjung scalp for critical analysis. The scientists unanimously agreed that the Khumjung scalp was a man-made artifact,
constructed from the skin of the Himalayan serow. Scientists in New
York, Paris, and London agreed.

Once I reached Chicago, I stayed there, as it would soon be Christ-
mas. I was home, my work completed. Hillary, Doig, and Chumbi,
however, took the scalp on to New York, Paris, and London and then
back to New Delhi. They flew to Kathmandu, arriving there two days
before the scalp was due to be delivered to Khumjung. The deadline
would have been missed without help from the Cook Electric Com-

pany of Chicago, which was installing a communication system in Nepal. We had met some of its officials there previously and they had generously offered to let Hillary fly the scalp back in their high-altitude helicopter. The day after they arrived at Kathmandu, Hillary, Doig, and Chumbi boarded the helicopter and started for Khumjung. Before long, they ran into heavy clouds and overcast and had to return to Kathmandu. On the very last day of our six-week permit they again boarded the helicopter and climbed skyward. The gods were smiling. They landed the helicopter in a field right near the village and returned the scalp in the nick of time. As an added dividend for the villagers' kindness in allowing us to take the scalp, Hillary promised to educate one of their bright young boys. A boy named Kalden from Khumjung was soon in Father Moran's school in Kathmandu.

In June 1961 I rejoined the expedition briefly in Kathmandu. There had been no new developments regarding the Abominable Snowman. Carol was with me on this trip, and after my work was completed at Kathmandu, we went round the world on our delayed honeymoon.

After the expedition was completed, Hillary and Doig wrote *High in the Cold, Thin Air*. My report on the Abominable Snowman appeared in the 1962 yearbook of the *World Book Encyclopedia*. I had a lecture film, a slide talk, and hundreds of slides to remind me of this glorious expedition. It was indeed one of the high points of my life, and I came away with the greatest admiration for Sir Edmund Hillary, Desmond Doig, and all those dedicated men who made the expedition a success.

Even though I feel sure that each bit of evidence for the Abominable Snowman had a natural explanation, I know that belief in the yeti will continue among the Sherpas and the Tibetans and many other people who like to think there is a creature not yet discovered. And so, despite my personal feelings, which are a result of our investigation, I suspect that the yeti will live on in the minds of many people. I would have liked to have found the yeti and brought home a good photograph or the living animal itself. I can fully understand the thinking of those who believe there is still a yeti up there waiting to be discovered.

17

"Wild Kingdom" and a Railroad in a Zoo

Don Meier showed the pilot film for "Wild Kingdom" to various agencies and prospective clients, and several expressed considerable interest. Meanwhile, in 1962 I had accepted an invitation to speak at the Zoological Society of Omaha; the society had recently received a large bequest and wanted to develop the city zoo. At the banquet I sat next to V. J. Skutt, chairman of the board of Mutual of Omaha. He asked me what was new in television, and I told him about the pilot film. He showed immediate interest and wanted to talk about it in detail. He asked when I had to leave Omaha; I said the next day. "Okay," he said, "what are you doing for breakfast?"

I said I'd be glad to have breakfast with him and invited him to the hotel.

"No," he said, "come down to our office and we'll have coffee and doughnuts there."

The following morning in the dining room at Mutual of Omaha, we had breakfast with Meade Chamberlin, vice-president for advertising, and a number of other company officials. I talked about the pilot film and our plans for developing it into a series. Briefly outlining the project, I suggested they talk to Don Meier and urged them to view the pilot film. They were eager to do so. One thing they impressed upon me, however, was that in the past I had refused to do commercials for Mutual of Omaha, thinking that as a professional zoologist I would be doing them a disservice. They had a different concept. They wanted to integrate Mutual of Omaha into the program itself, and felt that my doing the commercials would facilitate this. In the following days I examined all of the various aspects of the company's handling of insurance, as well as its extensive facilities. I needed to make sure that I knew what I was talking about; I wanted to believe in it. To this day I am glad I did, for I was soon to discover that Mutual of Omaha was permeated with a feeling of service to others. As soon as I felt comfortable with the integrity of the Mutual of Omaha people, it was easy to do the commercials. Our relationship has been a long and happy one.

Don Meier phoned Chamberlin and arranged to show the pilot film in Omaha. As a result, and contingent upon getting the show on NBC, a contract was developed between Mutual of Omaha and Don Meier Productions. Mutual of Omaha negotiated with NBC for a good time on Sunday evening, and a contract was signed for thirteen episodes. Don Meier negotiated a contract with the Chicago Park District for the right to use their animals and my services. Then we started producing the thirteen episodes.

Jim Fowler took part in each program and became associate host. Jim was not a full-time employee, as he made his living traveling the lecture circuit with his birds of prey. I too worked only part time, for as director of the zoo I could get away for only two or three weeks at a time. Don Meier arranged traveling schedules accordingly, and Jim and I were on location sometimes together and sometimes separately.

The name of the program became "Mutual of Omaha's Wild Kingdom." Some of the footage in that first series was previously unaired film from the NBC-sponsored expedition to the Amazon.

"Wild Kingdom" started with the "Zooparade" background. For most of its eight-year run, "Zooparade" had been produced live. Only a few shows had been filmed—in Florida, Wyoming, East Africa, South Africa, and the Amazon River jungle. We had some 300 show ideas before we even started filming "Wild Kingdom," and one of those became the pilot film showing Jim Fowler and his eagles, hawks, and falcons. Others of the first series of thirteen episodes were made in a studio with zoo animals. The first episode, entitled "Design for Survival," investigated and demonstrated a wide range of protective devices used by wild creatures: camouflage, protective armor (illustrated with an armadillo), mimicry, the feigning of death, bluffing, and the use of claws, teeth, horns, antlers, and poison. Other programs of the first series dealt with methods, both primitive and modern, used to capture wild animals; myths and superstitions about animals; strange courtship and eating habits among wild creatures; and the use of sonarlike sound emissions among such animals as bats. Another program examined the big cats of the world—lions, tigers, jaguars—their intelligence, their weapons, their agility, and their instinctive behavior. The series appeared to be well received, as indicated by high ratings and an encouraging volume of mail from viewers.

During the early summer of 1962, Howard Baer, president of the Zoological Board of Control of Saint Louis, came to see me in Chicago. George Vierheller had retired in April and the board was looking for a zoo director. I went to Saint Louis for further talks with Baer and other board members. They seemed definitely interested in me. I told them I'd like a few days to consider and would let them know.

Back in Chicago, I told the general superintendent and the director of the division of special services of the Chicago Park District of the Saint Louis offer and asked for their reaction. They consulted the other board members, but could not match Saint Louis' offer. I therefore agreed to go to Saint Louis toward the end of September.

Don Meier worked out an agreement with the Zoological Board of Control to allow for my television work, for which he would pay the zoo $25,000 per year. This, in effect, gave the zoo my services free, plus the benefit of a television program mentioning the Saint Louis Zoo in each episode. In return, the zoo would allow use of its animals and my absence from the zoo when I was filming in various parts of the world.

I had spent eighteen years in Chicago as director of the Lincoln Park Zoo. I was fond of the city and pleased with the facilities of the Chicago Park District, though somewhat disappointed that they were not interested in capital improvements at the zoo. I had persuaded them to build a children's zoo and a farm-in-the-zoo, and to remodel a few buildings, such as the bird house. But plans for other buildings were always shelved and funds for development were never available. Pay was low for the director, but high for the animal keepers and other personnel. I was happy to have the animal keepers upgraded, and continued to work to change their status from laborer to skilled technician.

It was an upheaval for me to leave Chicago, where I had lived for so long. On the other hand, Saint Louis was home, the place where I had started my zoo career. I liked the city and knew a lot of people there. And I liked the zoo—liked its beautiful location on eighty-three acres in Forest Park and admired the modern concepts George Vierheller and John Wallace had developed.

The zoo's basic income came from the city of Saint Louis, which now gave three mills per dollar of city taxes to the zoo, to be administered through the Zoological Board of Control. This method of financing removed the zoo from politics and spared it from competition with the essential services for budget money. Those three mills on the dollar could be spent only for the zoo, and they could be accumulated from one year to the next if need be.

I had been at the Saint Louis Zoo for the construction of the reptile house, the bird house, the antelope complex, and the small-mammal pits; I knew it was a zoo with a future. Howard Baer was honest with me. He told me that while the zoo was getting along on the money it was receiving, and still had some bond money for capital expenditures, funds would soon be exhausted and additional sources of income would have to be found in the near future. That was a challenge. But I knew the great love the people in Saint Louis had for their zoo.

Carol and I went to Saint Louis to look for a place to live. We wanted to be near the zoo and we also wanted an area with good schools and nice neighbors. We settled on a small house in the Hillcrest district of Clayton, within walking distance of Washington University. Marguerite could attend the De Mun School and Fred, Clayton High School. Alice went back to Connecticut College for Women.

The thirteen episodes of "Mutual of Omaha's Wild Kingdom" started in January of 1963. Don Meier Productions kept its office in Chicago, but came to Saint Louis with a crew to do the studio portions at the Premier Studio. This was convenient for me, since the studio was only ten minutes from the zoo and, in case of an emergency, I could get back quickly.

We did some programs at the zoo, but after they had been aired, it was decided that all future programs would be in the Wild Kingdom. We continued to build new shows, and I would fly to Florida or Catalina or whatever location, film for two or three weeks, and then return to the zoo for a period.

I began to build a staff at the zoo. Moody was still there as general curator. His son, Jerry Lentz, started as an animal keeper and progressed rapidly. Bob Frueh also started as an animal keeper and went on to become curator of mammals. Charlie Hoessle, now the director, started as a keeper and graduated to the post of curator of reptiles. Mike Fleege, also a keeper at first, became curator of birds. Regular staff meetings were held in my office, and everybody had an opportunity to find out what was going on all over the zoo. Later on, another curator of mammals, Roger Birkel, was chosen to accompany me on a "Wild Kingdom" filming trip to South Africa.

Early in my directorship, a large parking lot was built and the zoo was fenced in. The fence stopped much of the vandalism in the zoo and gave all of us a greater feeling of security when we went home at night.

I consulted Bob Heath, a specialist in miniature railroads, about a transportation system for the zoo. The route was surveyed and pronounced feasible with the addition of two tunnels. Bob came up with a concept for financing it privately. A relative of his, Bob Murch of the Murch Construction Company of Saint Louis, became interested, as did Harry Batt of New Orleans, who owned and operated an amusement park. They formed a company and put up the money to build the railroad. A contract was signed between the zoo and the railroad company, and work began on a transportation system. The Murch Construction Company built the railroad and Bob and Shirley Murch operated it for the first three years.

Three trains were purchased. Each engine pulled six cars, each with open seats and a candy-striped canvas canopy over the top. Nearly ninety persons could be carried in each train. Retired railroad engineers were hired to run the trains and high-school and college boys to be ushers and conductors. Each train was fitted with a public-address system over which was played a tape-recorded description of the areas the train was passing. The tape also gave information about the next stop. There were four stations at first; a fifth was added when the children's zoo was completed.

The acceptance of this railroad was fantastic. During the first year, we carried more than a million passengers, a figure far beyond our wildest expectations. The line became known as the best miniature railroad operation in any zoo in the United States, and it was certainly the greatest money maker. This was true because it was a transportation system rather than just an amusement ride.

Construction of a central plaza at the zoo and the reconstruction of the waterfowl ponds were undertaken and quickly completed. These structures greatly improved the appearance and efficiency of that part of the zoo and facilitated the flow of traffic to the bear pits and eventually to the children's zoo, the aquatic house, and the elephant house.

The master plan that had been developed for the zoo before I became director called for destruction of the 1904 World's Fair Birdcage and the building of a similar cage on the south side of the grounds. That was a controversial subject. I felt that the old structure was a historic building, as it was the first walk-through birdcage in the world. And it is still the largest. It was built as the Smithsonian Institution exhibit for the 1904 World's Fair. In the 1960s I recommended that the cage be remodeled and improved. It was in a state of disrepair, but the engineers pronounced the beams solid and recommended only a little extra welding at the sidewalk level. The old wire was replaced with sparrow-proof, vinyl-coated wire that would never need to be painted. The concept of the walk-through birdcage was altered with an elevated walkway and a complete revamping of the ground level. We installed new trees and shrubs and a large pool with a waterfall which formed a stream through the whole cage, was filtered in the keepers' quarters underneath the sidewalk, and recirculated. Vestibules were built at both ends and curtained with hanging chains of plastic so that people could pass through but birds were discouraged from getting out.

The Niemeyer statue of the Zuni Indian with doves was set in a lovely little alcove at the east end of this birdcage.

It is interesting to note that when that cage was constructed in 1904 it cost the Smithsonian $15,000. In 1963 tearing it down would have cost $35,000 and duplicating it, $750,000. This shows what has happened to our money.

We quickly developed an educational department at the zoo with funds provided by a grant from the Greensfelder Foundation.

We remodeled a section of the basement of the primate house to make one large room with a secondary room for office space, library, and equipment. Charlie Hoessle was in charge of the educational department, in addition to his duties as curator of reptiles, and all the staff at the zoo participated.

We told the Board of Education of our plans, and the board appointed a coordinator who kept the schools and teachers informed about using this department at the zoo and working it into their classroom curricula. We gave a several-week course to teachers who wished to take advantage of this opportunity, and they received credit for it.

The zoo had a part-time veterinarian who was available on call. When Barry Commoner was establishing the Center for the Environment at Washington University, he asked the zoo to participate, and that association supplied funds for employing a full-time veterinarian-pathologist and setting up a tissue bank to preserve various tissues and organs of animals that died. Eventually, we published a catalog of these specimens which was available to schools or researchers anywhere for a nominal fee. With the catalogue went a census of the animals in our collection and a note advising qualified researchers that if tissues or organs from any of these animals were required, we would preserve or freeze them according to their instructions and make them available when the animal died.

The old small-mammal house that had been tacked onto the reptile house was a building of very small cages, hard to clean and not very uplifting to visit. We remodeled the building and created a zoo nursery where baby animals born at the zoo could be raised on formula behind plate-glass windows through which the public could view them. This was a training place for our personnel in preparation for the children's zoo to come.

The diets in the zoo were improved with the addition of a twenty-foot-square building where oat grass was grown hydroponically. To that building we added glass windows so that the public could look in and, as the label suggested, Watch the Grass Grow. This little building was completely air-conditioned to maintain a constant sixty-eight degrees, winter and summer. This temperature kept down the formation of mold. Chemicals were added to the water in the top trays and filtered down through seven other trays. The oat seeds were soaked overnight in water and then planted on wire mesh an inch above the floor of the tray on thin sheets of specially designed paper. The chemically treated water flowed down to fill the first row of trays. The flow was timed so that the seeds had about four minutes of submersion. Then the water drained down to the next tray, and so on down all seven trays and finally was spilled into a sewer at the bottom. Charts were posted showing the age of the grass in each tray—one day, two days, and so on. At the end of the seventh day, the oat grass was approximately ten inches tall. It was then harvested just before it started to joint—the optimum time for nutrition and tenderness.

We found we could harvest one ton of oat grass each day in this building. We fed it to all the grass-eating animals in the zoo: birds, large mammals, antelopes, deer, small mammals, lizards, ducks, and geese. It was a food supplement, not a total diet, but green grass contains elements that dry grass does not and is beneficial to animals. We found the zoo did not require a full ton of grass per day, so we cut production to one-half ton. If more was needed, production could be increased.

The zoo had three refreshment stands when I returned in 1962: the old Chinese pagoda stand at the top of the hill near the lion house, a new stand that had been erected on a main walkway in the western section (eventually next door to the children's zoo), and Zoo Central at the main crossroads by the sea-lion basin. These refreshment stands also sold souvenirs and provided a considerable part of the income of the zoo. In fact, the zoo could not operate without that income, and we tried everything we could to increase the revenue. For example, we built small stands in different locations and at busy times over the weekends set up temporary stands in the parking lots near the entrance gate.

The zoo had for years bought its ice and kept it in an ice house in back of Zoo Central. Our ice bill kept climbing and finally reached $9,000 per year. I then investigated ice-making machines and purchased one for $27,000 that could make more ice than the zoo could use. The machine paid for itself in three years; from then on the ice cost us only electricity, maintenance, and water.

On the three mills per dollar of city taxation that the zoo was receiving, our budget was tight. Howard Baer and I made several trips to Jefferson City and talked to the members of the legislature. A bill increasing the zoo's allotment to five mills was drawn up and passed. This eased our situation a little, but we still lacked money for capital expenditures. When additional funds were needed to build the children's zoo, the future earnings of the railroad were used as security for the zoo association to borrow the money.

A communication system was developed with walkie-talkies and radios in cars and trucks to maintain contact with Zoo Central as the base station. When a refreshment stand needed more ice, it could call the base station and in a short time ice would be delivered. The communications system speeded the security department's response to emergency calls and allowed people to be located easily when they were moving about on the zoo grounds.

Chimp shows continued to be popular. The Saint Louis Zoo had many famous chimps, including Mike, Henry, Duffy, Sammy, and Billy. When the chimp show got larger and was composed of ten to twelve chimps, one was always star of the show. One star was a magnificent physical specimen named Jimmy. He had come to the zoo in the mid-1930s as a tiny fellow weighing not more than thirty pounds. He started in the kindergarten, progressed through the ranks, became a star, and retained that position until he was twelve or thirteen years old. As he matured and got bigger and bigger, he began to assert his independence and had to be retired. He had a luxuriant coat of long black hair and was a very large chimpanzee. Eventually he was sold to the New York Zoological Park for their new anthropoid ape house.

Years afterward, on a visit to the Bronx Zoo, Lee Crandall, who was taking me around, came to the ape house and said, "Marlin, we have an old friend of yours in here. Come take a look." I walked over in front of the cage and there was one of the biggest chimps I'd ever seen. I thought and thought, but I couldn't place him. Lee said, "Why, Marlin, that's Jimmy from the Saint Louis Zoo." It had been twenty years since I had seen him. "I remember how expert Jimmy became at walking on his hands," I said, "and he used to give a kind of greeting with his lips and tongue." I stepped over the guardrail and spoke to the chimp. He immediately came over and gave me this greeting with peculiar lip and tongue action. Then I indicated to him to walk on his hands, and he did! He then came over and held out his hand as though he recognized me—and I think he did.

Most chimps developed just like Jimmy at the Saint Louis Zoo, starting in kindergarten and working their way up. But one star of the chimpanzee show didn't. He had already had some training when

George Vierheller bought him in 1960. He belonged to a young man in Florida who had taught him to say "Mama" and had named him Mr. Moke. He was small enough so that Mike Kostial could put him into the training section for the show, and he learned many interesting tricks. Mike would set up a blackboard facing the audience, lead Mr. Moke to it, then pick up a microphone while Mr. Moke picked up a piece of chalk. Mike would explain that Mr. Moke could write. At a suggestion from Mike, Mr. Moke would write an *M*, then an *A*, another *M*, and another *A*. Then he would walk over to Mike and take his hand; Mike would lean over and say, "Now, Mr. Moke, you tell all the children out there what you have written on the blackboard." Mr. Moke would say "Mama" with a throaty sound, but distinctly enough so that anybody could understand.

When I returned to the Saint Louis Zoo in 1962 and "Wild Kingdom" was filming its openings and closings at the Premier Studios, Mr. Moke became part of the show and was there to demonstrate something of the learning power of chimpanzees. We had him putting square pegs into square holes, round pegs into round holes, and triangular pegs into triangular holes. He did this very well.

As the years passed, Mr. Moke grew and eventually matured. Mike Kostial, of course, was aware of this, as was everybody else at the zoo. But Mr. Moke was still tractable and was still being used in TV work. One day he was scheduled to be at the studio. Mike and Henry Turnis accompanied him. Mr. Moke came over to me and Jim Fowler as usual. After a take or two, Mr. Moke suddenly went berserk! He pulled away from Mike and ran around the studio, slapping his feet against the floor, an indication of aggression, and getting ready for an attack—hair upraised, threatening position, grabbing things when he went by, turning over tables and light stands. Then he got behind the set into a storage area and started throwing things around. This went on for nearly an hour. Mike had experienced such behavior with other chimps and knew that we couldn't force Mr. Moke to stop. We just had to wait until his mood changed. Eventually that happened; Mr. Moke walked over to Mike, held out his hand, and said, *Ooh, ooh, ooh, ooh,* to let Mike know that he was a friend again.

That was Mr. Moke's last appearance on "Wild Kingdom," but Mike continued to use him in the chimp show until about ten days later, he went berserk again right in the middle of the show. He was threatening and set a bad example for the other chimps. So Mr. Moke was retired.

Just before I returned to the zoo in 1962, assistant director Henry Sanders traveled to Europe to visit various zoos and to go to Hamburg, where he was to inspect and, if in good condition, bring back with him a young walrus that George Vierheller had purchased from Carl Hagenbeck. On his way to Hamburg, Henry stopped in Paris, went to see the two zoos, and was escorted around by Dr. Novel, the director. Dr. Novel showed him a chimp that had just arrived from French Cameroons and asked if the Saint Louis Zoo wanted it. Knowing that the zoo was always interested in a potential star for the chimp show, Henry looked at the chimp and played with him a little and thought he would be a good prospect. He was well proportioned, healthy, had good hair and intelligent eyes, and was very friendly. He was an appealing animal. Henry discussed with Dr. Novel the possibility of an exchange.

"We are very low on North American reptiles," Dr. Novel said. "We'd like some rattlesnakes and copperheads and various nonpoisonous snakes."

Henry closed the deal quickly.

Henry finished the trip by coming back with the walrus, which he named Siegfried. Siegfried was established in the new aquatic house, and there he made a tremendous hit. He was tame and gentle and liked people and liked to have his whiskers rubbed. He grew and grew and finally got to be absolutely enormous. Even then, in his outside pool, you could slap your hand on the wall and Siegfried would come up and let you rub his whiskers.

The little chimp from the Paris Zoo arrived, and being from France, was named Little Pierre. Mike started him in the prekindergarten, and he was certainly one of the most appealing chimps any of us had ever seen. Carol wrote a book about how Little Pierre came to the zoo and learned to be part of the chimp show. The royalties she made she gave to the zoo to be used for tables and chairs in the employees' lunchroom.

When Little Pierre was still small, Mike rigged up a cable that ran from the stage to the roof at the back of the arena. He gave a build-up about how the new arrival had recently flown in from Paris. Then he said, "Ah, and here he comes now!" In a little basket under a pulley, on the cable, came Little Pierre being pulled down by a piece of clothesline. Mike reached up and took the chimp out of the basket. Pierre learned to walk a tightrope and do all kinds of acrobatic stunts and won everyone's heart.

After Mr. Moke's episode on the set of "Wild Kingdom," we brought Little Pierre to the studio to take his place. I used to open or close the show with the little chimp sitting on my lap. He would look up at me intently as though he was listening to every word. Instead of calling him Little Pierre, however, we changed his name just for television. He became W. K.—which, of course, is short for "Wild Kingdom."

18

With the Bushmen
of the Kalahari Desert

While filming for the second season of "Mutual of Omaha's Wild Kingdom," Carol and I flew to Jan Smuts Airport near Johannesburg, South Africa, and joined our photographer, Warren Garst, and his wife Genny at their hotel in Pretoria. Most of the streets of this interesting capital city were lined with jacaranda trees and the trees were in full bloom. They formed a corridor of lavender, and as we looked down our sloping street to the downtown section and followed it up another slope beyond, we thought we had never seen a more glorious sight.

Warren and Genny had been busy getting gear and equipment ready for a trip to the Kalahari Desert. After another day spent buying the food we would need, the rented Land Rover was packed and we were off. A good tarmac road took us away from Pretoria through a country of changing scenery—mountains, plains, farmland, orchards, an occasional river. Eventually the paved road ended, and we drove over gravel and dust the rest of the way to the town of Kuruman.

Kuruman is an oasis in a desert. A large spring there forms a lake in the small town, and the surrounding country is dry scrub rather than true desert. We pulled into a service station for a rest stop, and the friendly man there asked us where we were from and where we were going, and whether we had seen the church. We had not, and asked, "What church?"

"Our Moffat's church, built long ago by missionary Robert Moffat. It's well worth seeing, as today it is just as it was when it was built over a century ago. That is where Livingston came when he first arrived in Africa, and it was Moffat's daughter Mary he wed."

It was indeed a charming little stone church, made in the form of a cross. The caretaker let us in and explained that the floor had never been changed from the day the church was first opened. The floor was made of clay mixed with the droppings of oxen. It had, of course, been damp at first, but after it had dried and was hard, it had been polished. The caretaker said all the early settlers had made their floors that way. It looked and felt like tile or polished concrete. I took pictures of the church, both inside and out. We did not know it then, but this contact with Livingstone was to kindle a spark in Carol that would take us to several other places he had visited in Africa on his travels in search for the origin of the Nile, including Ujiji on the edge of Lake Tanganyika, where Stanley found him. Carol eventually amassed enough slides to illustrate a fine talk about Livingston, which she has given to many audiences.

Late that afternoon we arrived at the Gemsbok National Park in the southwestern Kalahari. We were to stay in two cottages between which there was an outdoor tap for water, which was rich in minerals, including sulfur. The taste and odor were pungent, but at least it was water. We made coffee, but because of the water it didn't taste like coffee. We then made tea, but that didn't taste like tea either.

Next morning we arranged with officials at Gemsbok National Park to help us film a story about how the Bushmen live. The Bushmen are only about five feet tall, with strong, brown-skinned bodies. They make their few clothes of animal skins, grass, and leaves, but they do not wear many clothes. We first filmed the gathering of grass and branches to build a hut. Sticks were bent, and each end was stuck into the ground a foot away from another bent stick that crossed the first at the top. Other sticks were added until a round, domed structure about five feet high took shape. A door was fashioned, and the whole hut was covered with bundles of grass, starting at the ground and overlapping, like thatching, so that if it rained at least some of the water would run off.

The Bushmen are hunters and gatherers. The women and children gather roots, berries, and other edible plants, including melons. The men stalk and shoot with their small bows and arrows such animals as gemsbok, other antelopes, rabbits and ground squirrels, francolins, and guinea fowl; when chance permits, they take eggs from ostrich nests. They drill an ostrich egg at one end and collect the contents in a pan by blowing through a straw inserted through the hole. The eggshell is saved for use as a canteen for water. A tuft of hard grass makes a stopper, and the water-filled egg is carried in a net made of grass twisted into string. They sometimes bury the eggs in the sand as a cache for a return from a long trip.

Bushmen have great stamina. After shooting an antelope they can pursue it for hours until it is so weakened by the arrow and loss of blood that they can kill it. Virtually no part of the animal is wasted. The liquid from the stomach is drunk to allay the thirst from the long run. The liver is eaten raw. The animal is cut into sections so it can be carried on the hunters' shoulders back to camp.

After the Bushmen had feasted on such an animal, we were able to film the celebration dance that followed. The dancers formed a line and followed each other in a simple, shuffling dance around a small circle to a tune made by a wooden bow held to the teeth and a string rubbed with a stick. The effect was similar to a violin, while the tones were changed by the mouth in the manner of a mouth harp. This music was accompanied by sing-song and grunts in time with the shuffling of the feet.

We filmed the Bushmen getting ready for this dance, decorating their faces and bodies with white, black, and red designs.

The young boys showed us a game they played. A little mound of grass about a foot high was made on the ground, and about fifty feet away another mound was constructed. A small stick about three feet long was grasped at one end and thrown at the near mound so that it hit with a flat, glancing blow, bounced upward, and flew through the air a distance of sixty or seventy feet. After three or four boys had thrown their sticks, they ran to them, and starting at the second mound bounced them back to the first. There was much excitement as the players gathered the sticks after a throw to see whose stick had gone the farthest. I tried the game, but the kids always beat me.

One afternoon we returned to our cabins as a dark cloud was threatening. When the rain struck, sheets of water ran off the roofs of our cabins, and we all dashed outside with pots and pans, pitchers, and a wash basin to catch the fresh water. We stood in the deluge with mouths open, drinking the fresh water we had been starving for. This would make coffee and tea taste as they should. We filled everything we had that would hold water. Even after Warren mentioned that he had noticed bird droppings on our roof, our high spirits were not dampened, for now we did not have to drink water that smelled like rotten eggs.

After the shower stopped, Carol and I changed to dry clothes and went for a walk to the red sand dunes not far away. When we got there, we found a whole group of naked children on the ridge of a long dune, jumping and rolling down the wet sand. They were having so much fun that I ran to get Warren and my own camera. Warren filmed a good series for the Bushman show, and I was pleased with the pictures I took.

Carol was so interested in the Bushmen that when she got home she joined forces with one in writing a story about them that was published by Atheneum as *I Saw You from Afar*. The title was taken from the usual salutation of the Bushmen. Being small people, they compliment a visitor by saying, "I saw you from afar," implying that, being tall (important), he was easily seen from a long distance.

19

From a Rhodesian Waterhole to the Great Barrier Reef

"Wild Kingdom" was renewed year after year, and its personnel continued to travel. I was happy that Carol could go with me on many of these trips. It was always much more fun when she was along.

On one such trip, she went camping with us at Kennedy II waterhole in the Wankie Game Reserve in Rhodesia when we were filming the activities of the conservation department capturing a young giraffe to be transferred to the new sanctuary at Kyle. Our camp was pitched within sight of the waterhole and any time of the day or night we could see animals coming to drink. It was like having box seats at the opera. We saw buffaloes, elephants, giraffes, zebras, and the beautiful sable antelopes. Sometimes the sables would come in as a group; sometimes just one lone male would arrive with the other animal species. We saw lions, hyenas and jackals, baboons, monkeys, and monitor lizards. After a few days of watching, we realized that the giraffes came to drink at four in the afternoon. We began expecting them.

Don Meier decided to film us riding near the giraffes. The cameras were set up, and Jim Fowler and I, with some of the game department rangers, drove in our Land Rover along the other side of the waterhole. As we moved into the picture, the giraffes moved away from us—all except one male, who stood watching us. Realizing there was something wrong with him, we backed off to get further away from him. And then we noticed a lone female that had run a short distance, had stopped, and was looking back. The male started walking slowly (he had some kind of leg injury) to the female, and they continued on together. I was touched to realize that animals have concern for each other and will wait to see if some help can be given.

Carol had traveled over much of Africa with me and knew it well. Often friends had asked if they could go with us on our filming trips, but of course this was impossible because of the tight schedules, travel arrangements to reach remote places, and lack of accommodations for any extra people. On one occasion, however, after our children were away in school and I was filming somewhere where it was not possible even for Carol to be with me, she decided that rather than be alone in Saint Louis for so many weeks, she would lead her own safari to the places in East Africa she loved best. She gathered up fifteen of her friends and arranged with old friends in Nairobi to take them all on a journey through Kenya, Uganda, and Tanzania. They were gone for a month and had a wonderful time. Many of these same friends and others besides have since traveled under Carol's leadership to Africa and other parts of the world that Carol first came to know in her travels with me.

When Carol learned I was going to Australia to film a program on the Great Barrier Reef, she organized a trip to the South Pacific, New Zealand, and Australia and timed it so we could be there together. When the filming crew had reached a point where they could continue without me, I boarded a plane and flew from Sydney to join Carol in New Zealand.

While in Aukland, we invited Sir Edmund Hillary and his wife, Lady Louise, to join us for dinner. It was wonderful to see Ed and Louise again and they charmed Carol's group. We saw the Glow Worm Caves, the lush farmsites of the North Island, and the geysers and hot springs and the mountains.

By the time Carol and her group were ready to leave for the South Island, I was due to rejoin the filming crew and finish my part in the sequences on the Great Barrier Reef.

Of all the coral reefs in the world, the Great Barrier Reef of Australia is the largest. It extends 1,500 miles north and south off the east coast of Australia. We have filmed there on several occasions. On one of those we worked at the reef's eastern extreme, offshore from Brisbane, at a place called Saumarez Reef. We had left our mother ship in a smaller boat and anchored over a likely looking spot. Six of us went into the water to film what we could find in that region. Poking around the nooks and crannies of the reef, we came to a spot like a glade in a forest, with white sand bottom. Working around its edge, we noticed a six-foot tuna swimming erratically. Something was wrong with it. It was ill. It settled at the bottom of the white sand glade and there, in a spasm of erratic muscular contortions, it expired. We thought this very strange and couldn't imagine what had happened to the fish. But we didn't need to wait long to find out what would be its end. Three gray reef sharks came zooming in. The first reached the tuna and bit its side, whirling and rotating its own body so that it cut out a bite-size chunk. It then swam away in a large circle while the second shark came in, followed by a third. Finally there were five gray reef sharks devouring that dead tuna. We filmed this, remaining motionless, watching from the so-called security of a coral head. We were not more than twenty feet from where the sharks were feeding and had a clear view of their amazing feeding methods and the way they rotated their own bodies in order to cut out sizable pieces from their prey. I had imagined a shark just bit with its mouth; instead, it would get a mouth hold, then swing its body in rotation to tear the piece loose. We were absolutely sure—at least I was absolutely sure—that those sharks were so interested in their food that they wouldn't bother us. I watched for some time. Then, as my gauge indicated my tank was low on air, I signaled the other divers and headed back to the boat. I started up on an angle, swimming above where the sharks were still feeding on the remains of the tuna. I aimed for the boat itself, the bottom of which I could see. About two-thirds of the way to the boat, I had a shivery feeling and glanced back to see what was following me. There *was* something following—the other divers.

Another trip to the Great Barrier Reef was designed to study and capture sea snakes, which in some sections were quite abundant. We were accompanied by the director of Marineland of Australia, who wanted specimens for exhibition. Before we dove in this section, John Reynolds, the director, cautioned us that there would be danger if the sea snakes spotted us and swam directly at us in what he called an "attack." All sea snakes are highly venomous; they are related to cobras and have a neurotoxic venom that can be deadly.

With me was Tom Allen, one of Ross Allen's sons. Tom had been raised with snakes at Ross Allen's Reptile Institute at Silver Springs, Florida. He knew them very well and had captured every kind of poisonous snake in the United States. I talked to Tom about the danger of an "attack." We both found it hard to believe that a sea snake would go out of his way to attack a person. He'd have no reason to do so, as he couldn't eat us, and we weren't threatening his territory. I had a feeling that the "attack" was simply an approach made out of curiosity. Tom agreed with me.

So when I was at the bottom of the ocean and saw a snake swimming toward me, I held still in the water. I sat down and waited for the snake to approach. He came first to my flipper and then to my ankle; he worked his way up to my knee, then to my arm. He was going to go a little farther, but I gently reached out with my other hand, took hold of his body without actually squeezing him, and moved him away from where he was headed. We did this on several occasions. As the sea snake was moving up my body, I could see that his tongue was going in and out and I had the impression he was actually smelling me. A snake uses its forked tongue to touch an object; it then withdraws the tongue into its mouth and inserts it into its nostrils from the inside of its mouth. In this way the snake touches its Jacobson's organ, its organ of smell. I think the sea snake was simply curious about this strange object (me) in the water. He was merely investigating and had no thought of biting or injecting his venom in either me or Tom. We found it was easy to put a hand in one of our long, knitted diver's sacks, take hold of the body of a snake through the sack and slowly pull, inverting the sack to the point where the snake was inside it. We caught quite a few sea snakes for Marineland of Australia on that occasion.

In that same location we saw the biggest moray eel I have ever encountered. He must have been six or seven feet long; his body was massive in girth and green in color. He was swimming slowly just over our heads while we were near the bottom. He was moving on an angle toward some coral. He entered a hole in the coral and disappeared; in a short time his head reappeared at the entrance. He'd gone in, turned around, and looked out from the entrance; we could see him breathing. Tom signaled that he was going up to the boat directly above us, and we knew immediately what he had in mind. He got some fresh fish from the boat and put them in one of our diving sacks. Then he came back down, showed the fish to the photographers, and indicated by sign language that he was going to offer a fish to the moray eel. The photographers took their positions and the cameras rolled. Tom inched over to the eel. As soon as the moray realized that here was something to eat, he became interested and moved out toward the fish. Tom backed up a little and the eel came further out. Tom kept moving back and the eel kept coming forward until finally Tom lured that eel clear out into the open from his hole in the coral. Tom gave him the fish; he took it, then turned back to his hole in the rock, swallowing the fish as he swam.

A second time Tom offered the moray eel a fish, and a second time lured him into the open from his coral fortress and then fed him. All this was recorded on film for "Wild Kingdom."

When the sequences in which I appeared had been shot on this latest visit to the Great Barrier Reef, we packed up and prepared to move to the next filming location, Kangaroo Island, offshore from Adelaide. There we would film the animals at the Flinders Chase preserve. Our stopover in Melbourne coincided with a visit by Carol and the group, so we all had dinner together. While Carol and her party attended the prestigious Melbourne Cup races, the "Wild Kingdom" crew flew off to Kangaroo Island.

A week or so later, I met Carol and the group in Sydney. Having completed my filming assignment in Australia, I joined them for the trip to Manangrida, a station where the government maintains schools and a center for the aborigines of the region.

A friend from our last trip, Reverend Gowan Armstrong, met us, along with the administrator of the center, William Clark. We spent a delightful day during which they had arranged an exhibition of special dances and craft activities by the aborigines. We bought bark paintings and pandanus mats and admired a didgeridoo, a six-foot-long section of a small tree, about four inches in diameter, hollowed out by termites. When they find such a tree, the aborigines cut off the hollow section, burn out the rest of the inside with glowing sticks, smooth off the outside at a place where they are going to put their lips for blowing, and paint the didgeridoo in colorful designs of their totems. It becomes a musical instrument used in dances and other ceremonies. Didgeridoos were in short supply at the time, so I couldn't buy one. Gowan Armstrong offered to ship one out later, and this was quite agreeable to me as the six-foot-long instrument would have been difficult to carry on the plane. When my didgeridoo was finally ready for shipment, Gowan found that the post office wouldn't accept such a long parcel. He had to ship it by sea freight. When at last it arrived in Saint Louis, we were aghast to find that the minimum charge for sea freight made my didgeridoo one of the most expensive artifacts in our home.

20

Tibetan Dogs and American Wolves

Carol planned a journey to India, Nepal, and Sikkim, and a group of our friends and our daughter Marguerite went along. I was able to accompany them to New Delhi, to Agra to see the Taj Mahal, and on to Kathmandu in Nepal before going to East Africa to film a story about flamingos on Lake Nakuru. In Kathmandu we looked up Boris Lissanevitch to find that he had closed the old Royal Hotel and had opened a dining room called the Yak and Yeti in another rajah's palace.

After a few days in Kathmandu I regretfully left Carol and Marguerite and flew off to Africa to join the "Wild Kingdom" crew. Carol and her group remained in Asia and visited Tiger Tops, where they rode elephants through the forest to the fabled and comfortable lodge deep in the terai. A sanctuary has been established there to protect the rare Indian rhino and the even rarer tiger. The group then traveled on to the hill station of Darjeeling in northern India. At an altitude of 7,000 feet, the Darjeeling area is a favorite resort for escaping the intense heat of summer. But the climate was really chilly in October when they arrived at the old Mount Everest Hotel.

The whole group had gone to the Tibetan Self-Help Center on

the outskirts of that small city to visit the Tibetan refugees and to buy some of the beautiful wool rugs that are made there. Carol and Marguerite never got as far as the rugs. They saw a little black-and-white puppy. It had been raining and the puppy was wet and dirty, but Marguerite picked it up and held it in her arms. It snuggled down into her arms and wagged its tail. It was love at first sight. When Marguerite asked to buy him, papers were produced declaring him a Tibetan spaniel named Tashi Dorja. The owners agreed to sell him, and included a homemade wooden box for him to ride in on the long journey home. Carol and Marguerite carried little Tashi on to Sikkim, over high and rugged mountain roads, and then back to Bagdogra, India, to catch a plane for Calcutta.

Carol had carefully checked at the United States Embassy in Nepal to see if there were any restrictions about taking a puppy out of India or bringing one into the United States. There were none, as no vaccinations are required for a dog under six months of age.

Because of Tashi Carol and Marguerite decided to come straight home while most of the group went on around the world. Everything went fine until they were due to leave the New Delhi airport, quite late at night, for a flight to Vienna. When they checked Tashi in for his "livestock" permit to go on the plane, the man behind the counter took the dog away from them and put him, in his box, back on a heap of empty cartons and trash. He told them they needed an export permit. Carol knew that this was not true, but no amount of arguments or pleading would budge the official. The plane was loaded and waiting for them to board. It had been a long, hot day of travel in India, and this was the last straw. They burst into tears of exhaustion and hopelessness.

At that moment a Swan Tour from England passed through the airport. When they saw the only other two non-Indians sobbing they hurried over to help. The airport official, seeing reinforcements arriving, gave up waiting for Carol to think of offering him a bribe and set his own price. He said that he could produce a permit for the equivalent of fifty dollars in rupees. He wrote something on a scrap of paper and gave it to them with their dog and put the money in his pocket. And so they brought Tashi home.

It was about three weeks later when I returned from Africa. Tashi was already in charge. Imagine my surprise when I started upstairs to find my way barred by a furiously barking little black-and-white dog.

A few people breed this rare dog in the United States, but the Tibetan spaniel is not yet recognized by the American Kennel Club. We had heard that there were some breeders of these dogs in England, so on a trip to London the next year, we carried a picture of Tashi with us. The Royal Kennel Club provided us with the name of Mrs. E.J.B. Wynyard, who has the Braeduke Tibetan Spaniels kennels in Northampton. We made an appointment, and armed with Tashi's picture, we rode the train out to Northampton, where Ann Wynyard met us with a Tibetan spaniel in her arms. She drove us to her beautiful country home; there we were greeted ecstatically by twenty-two dogs who all looked like Tashi except for their color. Mrs. Wynyard opened a window to the garden, and the dogs all streamed into her lovely living room but were soon ushered outside again. They were golden and sable and beige and white, but there wasn't a single black-and-white dog. She told us that particolored black and white was very rare in these dogs.

When we had assured her that we were reliable dog owners and she had marveled at Tashi's picture, we were able to buy a little golden female to take home as a gift to Marguerite—and to Tashi. We managed to get a veterinary certificate that same afternoon in Northampton and then Ann Wynyard took us back to the train for our return to London. When she placed the puppy in Carol's arms, her eyes filled with tears, then she turned and almost ran from the station. It is very difficult to part with a puppy.

When we arrived home in Saint Louis, Marguerite and Tashi were at the airport to meet us. We attracted quite a bit of attention with their joyous noises when they discovered the new dog. Marguerite named her Lhotse-La, after one of the great peaks in the Himalayas. Over the last nine years Tashi and his Lhotse-La have become the parents of six beautiful puppies. One female went back to England to Ann Wynyard and the other five live with us. We could not part with them and wonder how we ever lived without Tibetan spaniels. Tashi and Lhotse-La live with Marguerite and her husband, Peter Sorum, in Washington, D.C.

After her last trip to Australia, Carol wrote another book about the aborigines, *Boomerangs Returning*. She developed a series of slide lectures on the various trips and has shown these to civic groups, clubs, churches, and the Boy and Girl Scouts, not only in Saint Louis but in other cities as well. She put together a weekly television program called "Our Endangered World"—a conservation segment of the "Noon News" on KSD-TV, Channel 5. Each Monday she would report on some phase of conservation. When she was on safari to East Africa, she reported back by telephone, telling about the conservation problems of Kenya and Tanzania.

Carol also went onto the ice off the Magdalen Islands at the mouth of the Saint Lawrence River with Brian Davies to make a filmed report for her program on the cruel killing of baby harp seals. She became a strong voice for conservation in the Saint Louis area.

In 1961, after a meeting with Dr. Owen Sexton of the ecology department of Washington University, we and some interested friends organized the Wild Canid Survival and Research Center. This is a nonprofit organization chartered in Missouri with the objective of preserving wolves in as natural conditions as possible. I had long been concerned over the tragic decline in the wolf population of North America. This magnificent animal, the largest member of the dog family, was rapidly being exterminated and faced extinction. Harmless to man, the wolf has been the victim of baseless superstitions surviving from the Dark Ages. Its necessary and beneficial role in ecology—as a predator to eliminate the sick, weak, and excess young among large mammals—has generally been overlooked. In the United States today less than a thousand wolves exist, most of them in the Superior National Forest in northern Minnesota. A small group on Isle Royale in Michigan, formerly numbering about thirty, has been reduced to fewer than twenty in 1982.

We decided to set up our wolf sanctuary and study unit in the Tyson Park Research Center area near Saint Louis. The Tyson Research Center comprises 2,500 acres in rolling, wooded hills, about thirty minutes drive from Saint Louis. During World War II it was an army munitions dump, and it has bunkers and some old railroad tracks. It was and is well fenced around the entire perimeter.

After the war this reserve was made available to Washington University in Saint Louis as a biology center. The university and the director of the Tyson Research Center, Dr. Richard W. Coles, kindly let us have fifty acres near the center of this beautiful area for the wolves. It was important for us to have a secluded place for their protection.

We now have five major enclosures, about an acre and a half a piece, each with its own small lake, where the wolf packs may live as normal a life as possible while in our protective custody. They are fed Ralston Purina Dog Chow, with free choice of how much to eat and when, and twice a week as a special treat they are given chicken necks. In the wild a wolf that has made a kill will swallow the meal, then regurgitate part of it to feed the pups waiting in the den. The feedings of chicken necks maintain this natural behavior.

To head the Wild Canid Survival and Research Center, a board of directors was organized, consisting of prominent biologists, conservationists, authors, and sympathetic private citizens.

We have done a great deal of soul-searching over our decision to keep these wolves in captivity, in order to maintain gene pools of the most endangered subspecies and to keep these splendid animals in existence. They cannot be returned to the wild; at this time in our world there is no place for them to go. It will be a long time before people will unlearn all the false and frightening myths about wolves. It will be a long time before these intelligent animals will not be aggressively hunted from the air, on foot, and by snowmobile. The world's wolves are now nearly extinct. As man took over the land and brought his livestock with him, it was inevitable that a confrontation with natural predators would take place. There is no excuse, however, for man's invading the wolves' last, remote sanctuaries, where there is no competition, to kill those animals for so-called sport.

Since its founding ten years ago, the Wolf Sanctuary has established captive breeding programs with the aim of creating gene pools for endangered wolves. Of the three subspecies now living there, two of them are believed to be virtually extinct in the wild. At this time there are twenty-two wolves at the Sanctuary, which is supported totally by memberships and private donations. It has become a major wildlife resource and educational center, reaching more than ten thousand people a year. The Wolf Sanctuary also established a data bank for wolves in captivity throughout the world and maintains a strong educational program.

In 1974 the small group of Wolf Sanctuary volunteers sponsored the first International Symposium for Threatened and Endangered North American Wildlife, held in Washington, D.C. Every conservation organization was invited to participate. Every working biologist in the field came to report on his specialty: Roger Tory Peterson on birds, Margaret Owings on sea otters, Frank and John Craighead on grizzly bears, Charles Jonkel on polar bears, Brian Davies on harp seals, Michael Fox and Gordon Haber on wolves, Thomas Garrett on whales and sea mammals, F. Wayne King on reptiles and sea turtles, and many more.

Jimmy Stewart opened the meetings by reading Albert Schweitzer's "Reverence for Life" and Rogers Morton, Secretary of the Interior, gave the keynote address. The panels discussed many issues in wildlife survival, from habitats to captive breeding. Norman Baker, who had sailed with Thor Heyerdahl as navigator on the voyage of the reed ship *Ra*, reported on the pollution of the seas. He told how for days the sea was so dirty with flotsam and jetsam that they couldn't brush their teeth in its water.

Climax of the weekend symposium was a gala congressional dinner, so that the lawmakers could see how many people really care about the survival of our wild heritage. Our animals may live or die depending upon what our legislators decide in Washington. More than fifteen hundred people attended that landmark symposium.

In 1977 the Sanctuary sponsored another major symposium, this time in Saint Louis. We knew from the reports given in 1974 the approximate status of the various species. Now we were trying to find out what could be done about preserving them.

Carol keeps a poster pasted on our refrigerator reading: "Happy are those who dream dreams—and are ready to pay the price to make them come true." It is signed by Leo Cardinal Suenens, S.J.

Two lionesses, Amboseli National Park, Kenya. This is Africa to me—the vast distances, the lions, and the natural sounds of the animals and the wind. (*Photo by Marlin Perkins*)

(ABOVE) Cheetahs at Amboseli, with the 19,321-foot peak of Mount Kilimanjaro in the distance. (*Photo by Marlin Perkins*)

(ABOVE LEFT) Galloping wildebeests on their long migration across the Serengeti Plain. (*Photo by Marlin Perkins*)

(BELOW LEFT) Sable antelope herd, Shimba Hills Game Reserve, Kenya. Trophy hunting has relentlessly reduced these beautiful creatures to an endangered status. (*Photo by Carol Perkins*)

Walking the young giraffe back to camp in Wankie Game Reserve after its capture for relocation. (*Photo by Carol Perkins*)

"Wild Kingdom" camp in the Wankie Game Reserve. (*Photo by Carol Perkins*)

The bizarre hair-pulling ceremony in the Tecuna Indian village on the upper Amazon River. (*Photo by Marlin Perkins*)

Elephant seal chasing me out of his territory on an island off southern California. (*Photo by Warren Garst from Marlin Perkins Collection*)

In the Kalahari Desert I watched this Bushman draw designs on an
ostrich egg, which he later gave me. (*Marlin Perkins Collection*)

Displaying the Adventurers Club flag on the spectacular and difficult Tashi Lapcha Pass during our search in the Himalayas for the legendary yeti or Abominable Snowman. (*Photo by Marlin Perkins*)

r Edmund Hillary earing the alleged yeti alp in the Sherpa llage of Khumjung. *hoto by Marlin Perkins*)

We were fortunate to be at Mumadulai in India when twin baby elephants were born—a rare event among elephants. This photo was taken when the babies were less than one day old. (*Photo by Marlin Perkins*)

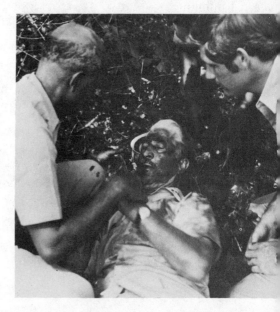

Just after the elephant flicked me out of his way. Forest Veterinary Officer V. Krishnamurti bathes my hand while Tom Allen watches to see what he can do to help. (*Photo by George Bokland*)

General Jimmy Doolittle (*left*) and V. J. Skutt, chairman of the boards of the Mutual of Omaha Companies, visit me on location at Mount Rushmore in South Dakota while filming an episode for "Wild Kingdom." (*Courtesy of Mutual of Omaha*)

Friendly and curious, some monkeys come close during filming to see what we are doing. (*Photo by Marlin Perkins*)

Humpback whale
breaching at Silver
Navidad Bank
in the Caribbean.
(*Photo by Marlin Perkins*)

Two humpback whales. (*Photo by Captain Dave Woodward*)

n Laysan Island in the
cific, getting ready to
orkle with monk seals.
*hoto by Peter Drowne from
arlin Perkins Collection*)

Examining a giant clam at Eniwetok Atoll. (*Photo by Rod Allin*)

Tashi, our first Tibetan spaniel, as a puppy.
(*Photo by Carol Perkins*)

Carol boarding our "yacht" for the long trip down Lake Tanganyika.
(*Photo by Marlin Perkins*)

My grandson, Peter Brentlinger, and I by a termite mound at
Polumnaruwa, Sri Lanka, with the top of a stupa in the background.
(*Photo by Peter Drowne from Marlin Perkins Collection*)

My daughter, Suzanne, during a family safari to Kenya, with her sons, Peter (*left*) and Chris Brentlinger. (*Photo by Marlin Perkins*)

(BELOW)
Our family on safari at Governor's Camp in the Masai Mara of Kenya in 1981. *Left to right, clockwise:* Ren Goltra, Alice Goltra, Peter Sorum, Carol, Fred Cotsworth, Mary Ellen Cotsworth, me, Marguerite Sorum. The occasion was Alice's birthday. (*Photo by Marlin Perkins*)

(ABOVE)
reindeer is harnessed and
the sledge is ready for the move
to the next camp during the
reindeer drive in Lapland.
(*Photo by Marlin Perkins*)

Woody, a four-month-old wolf
cub, at the Wild Canid Survival
and Research Center. (*Photo by
Jenny Silverstein from WCSRC*)

The "river truck" in which we traveled for five days on the Strickland River in Papua New Guinea. Peter Drowne forward, Carol and I midship, and Jerome Montague taking his turn driving. (*Marlin Perkins Collection*)

Tribesmen on the bank of the Strickland River watch us as we approach their village. (*Marlin Perkins Collection*)

Dick Johnson hand-feeding a white-tipped reef shark. (*Photo by Rod Allin*)

The gray reef shark attacking Dr. Don Nelson's submersible. (*Photo by Ralph Nelson*)

Measuring the largest dragon lizard we caught on the island of Komodo. It was seven feet in length. Dr. Walter Auffenberg kneeling. (*Marlin Perkins Collection*)

A dragon lizard makes its way along the beach of Komodo. (*Marlin Perkins Collection*)

21

Filming
the Wild Kingdom

About a year before it became time for me to retire from the Saint Louis Zoo, we brought in Bill Hoff as my replacement. Bill had started his zoo career under my direction at the Lincoln Park Zoo and had then become director of the Cincinnati Zoo. It was not easy for me to walk away from the zoo which had been such a vital part of my life for so many years, but I had long ago learned that nothing is permanent. Like a hyena puppy being weaned, I had protested and vocalized a little, but my sixty-fifth birthday was inevitably approaching. "Wild Kingdom" was continuing, and I was fortunate to have a second vocation.

In the early years of "Wild Kingdom" I had also been director of the Saint Louis Zoo, and had divided my time between zoo duties and filming expeditions. After retirement I could give more time to "Wild Kingdom," and so my time in the field was for some years increased. Now, however, only half my time is spent on "Wild Kingdom" location in remote parts of the world.

"Wild Kingdom" continues to be produced by Don Meier Productions in Chicago. The company is small, with Don as president and his wife, Lorie, as treasurer and accountant. Ten or twelve other people are employed at the office. Show ideas are gathered from many sources. Everyone connected with "Wild Kingdom" submits ideas, and over the years we have made friends with many conservation officials in the United States, Canada, and many other parts of the world. We're proud of the fact that we are often invited back to a filming location to record a second or third episode. The people we have worked with frequently alert us to new show ideas. These are investigated and if possible worked into the schedule.

Most of the logistic details for a show are arranged by the staff in Chicago before the crew is sent out into the field. The crew consists of one photographer (occasionally two photographers and very rarely three). A photographer on location becomes an associate producer; it is his job to contact all the people who will be involved, work out further local and logistic arrangements, and verify those that have been developed by correspondence and long-distance telephone calls from Chicago. Working with an outline determined earlier in Chicago, he amplifies and expands the story and forms his own concept of how the show should be put together. The photographers are all wild-life photographers, highly specialized in this field.

Warren Garst, our head photographer, who has been with us the longest, is also a zoologist. He taught zoology at the college level and filmed a true-life adventure series for Disney before he started working for "Wild Kingdom." He and his wife Genny are in the field much of the year, filming episodes. As many as twenty-four pieces of equipment are sent by air freight along with the photographer to the location site. After filming is completed, the exposed film is sent home through a shipping broker. It is never hand-carried or shipped as baggage. We have never lost a show because the film did not get back to Chicago.

In Chicago, the film is sent to a laboratory for development, and a color workprint is drawn. Don Meier views every inch of film that comes into that office and is frequently consulted by the film editors. If they run into problems, he helps sort out the difficulties. When the film has been edited to the proper length, it is shown to a writer who is given all the details: the people involved in the program, the locations, and the general story outline that was developed by the associate producer on location. The writer assembles a script consisting of no more than four words per foot of film. This is difficult writing and the reverse of Hollywood production. Hollywood writes a script and gives it to a producer, who gives it to a director. Then the scenes outlined in the script are filmed. We have to reverse that order because we're filming wild animals and can't predetermine what a wild animal will do.

Following the development of the script, I go to Chicago to record the narration. That's done in a sound studio. In addition, in a film studio I do the opening and closing and the introductions to each of the three segments of the show. Eventually, all the elements are put together; then prints are drawn, and the sound, including the sound effects tape-recorded on location, is incorporated.

Ten weeks is the minimum time needed, after the raw film reaches Chicago, to complete a program. Usually the interval is longer. Because of the uncertainty of filming wild animals, the shooting ratio for "Wild Kingdom" is as high as thirty or thirty-five to one. That means thirty feet of film are exposed to get one foot of film that will be used on the air.

Here is an outline of the equipment needed for the filming. On location we use two Arriflex cameras with 16-mm film and a variety of lenses. One of these cameras is available as a spare in case the other breaks down. Sometimes three are taken. We also take heavy wooden tripods in cylindrical tripod boxes. The Arriflex cameras are operated by batteries. We have both small, rectangular batteries and batteries worn on a belt around the camerman's waist; both types are attached by cord to the camera. Each camera is equipped with a film magazine and unexposed film is kept in specially designed cases. A box of essential tools is always at hand, as are transformers to convert 220- or 240-volt electrical circuits to the 110 volts needed to recharge the batteries. Another kind of transformer is used to recharge batteries in a location that lacks electrical service; this transformer can be used with an automobile or boat battery. A tape recorder is taken along and extra lengths of cord. We can synchronize the recorder with the cameras in order to record sound and action simultaneously. Or we can use the recorder alone to tape the sounds of animals or of vehicles—the starting of an automobile engine, the noise of an outboard motor on a boat, the sound of that motor fading in the distance. Such sounds are important in putting the program together. Over the years we have accumulated a large collection of animal sounds. These are all catalogued and available for use in case it's not possible to get a suitable recording on location.

We take lights if we're planning to film in subdued light or at night. And if underwater scenes are scheduled, we take scuba tanks and other diving gear, including a portable compressor to pump air into the tanks. And, of course, we take underwater cameras—Arriflex cameras that fit into a special underwater waterproof housing. Equipment for underwater photography is heavy and bulky and expensive to ship.

Our photographers also take Polaroid cameras, for two reasons. The first is to take pictures of every person working on camera in order to record the clothes he is wearing. As scenes are filmed over several days and out of sequence, it is important that a participant wear the same clothes every day he's on camera, and the Polaroid record helps us check that. The second reason we carry Polaroid cameras is to take instant pictures of local people working with us, particularly some of the more primitive natives in various parts of the world. A gift photo is often invaluable. It can often secure more cooperation and good-will than money could buy. Few people can resist photographs of themselves.

The fact that it is so hard to predict just what a large wild animal will do often causes a photographer working with such animals considerable worry over the safety of his motion-picture cameras, which are fragile, awkward and heavy to move, and not easily replaceable in the field. One time, when we were filming in Rhodesia (now Zimbabwe), part of our story was about the transfer of captured animals from the Wankie Game Reserve to a newly created reserve at Kyle. We started this episode by filming the white rhinos on the other side of Lake Kyle from the park headquarters. But one morning, when the game warden's duties kept him from working with us, we decided instead to film two giraffes that had been released in the park not far from headquarters.

Warren Garst and I drove out to look for them. We had been told that both giraffes were females, and that Mary, the larger one, had been hand-raised on a bottle and was still tame. We had no trouble finding the animals. As we topped a small rise in the road, there they were, across a large field of grass near some trees. As we wanted to shade our equipment from the hot sun, we parked the station wagon in that grove of small trees.

With Warren carrying the camera and tripod on his shoulder, we walked from the grove into the open field. Warren set up his tripod and looked through the viewfinder to compose the scene. "We're a little too far from them," he said. "Let's move closer. Then, when we have set up again, you can walk into the picture left to right, pause to look at the giraffe, and walk off in the same direction."

He was about to shoulder his camera and tripod again when we saw that Mary, the larger female, was walking toward us. We thought she would stop, so I took my position and waited. Warren was ready to start filming, and Mary walked right up to me, until she was towering over me. Then she lowered her head to my level. I raised my hand to let her smell me and get acquainted. That done, she lifted her head and walked over to Warren. She was now too close for pictures, so Warren gave her a chance to smell and get acquainted with him.

When Mary finally straightened up, we both thought she would walk away and allow us to film our scene. Instead, she became belligerent. She raised her front feet, one at a time, and struck the ground a fierce blow, very close to Warren. To avoid the blows, he quickly edged behind the tripod; I moved in at once and tried to shoo her away. But she continued her belligerence. Clearly, this was a disturbed and angry giraffe. I suggested to Warren that while I held her attention he should move away with the precious camera and tripod. I slowly maneuvered close to the only small tree in the open. There I picked up a branch for protection, and as soon as Warren had reached the grove where the station wagon was parked, I started backing away, threatening Mary with the branch whenever she came too close.

When I reached the dense grove of trees, Mary was at some disadvantage. And as soon as Warren was in the back of the station wagon with the camera and tripod, I slipped through the thickly growing trees to the driver's door and got in behind the wheel. Mary retreated to the open field when I started the engine. But no sooner had we cleared the grove than she was running alongside the car. I drove faster; she increased her pace. When she was running her fastest I checked the speedometer and clocked her at thirty-five miles an hour. We came to a fork in the road, and I turned left. For that I had to slow, but Mary didn't. She was still obviously upset. I doubled back onto the road leading to the grove; she was right there beside us. As she put on speed trying to get ahead of us, I braked quickly and cut across the lower end of the fork onto the road we had just left. Mary could not match that quick maneuver and we soon outdistanced her on the way back to camp.

When we told the game warden of our experience, he asked,

"Didn't you have a cookie for her? We always feed her when she comes to us."

I was reminded of the time our zoo keeper in Buffalo was badly injured because he didn't know that his assistant had been feeding the big deer hunks of bread. Because we didn't know what Mary expected of us, we almost lost some valuable camera equipment that day.

In 1982 "Mutual of Omaha's Wild Kingdom" celebrated its twentieth year of continuous production, one of the longest runs in television history. The ratings have continued to grow until we have reached an audience of 30 to 35 million people during each week of the heavy winter viewing season. During the past season the show was carried on approximately 220 TV stations in the United States and Canada, as well as stations in many foreign countries.

All of us who work on the program are very proud of the awards that have been presented to "Wild Kingdom" since it became standard viewing on the NBC network. They were awarded to the show, to Mutual of Omaha, to Don Meier as producer, and to me. This was gratifying acclaim, and Mutual of Omaha was proud to be able to announce in its publicity that "Wild Kingdom" has been honored on four separate occasions with the Emmy Award of the National Academy of Television Arts and Sciences. We have also received the first annual Communications Award of the National Wildlife Federation, a special Golden Mike award from the American Legion in Washington, and a number of other awards and citations. A study conducted by Stephen Kellert of Yale University showed that approximately 61 percent of the television viewers questioned reported "Wild Kingdom" as having had a very strong or at least moderate influence on their views on and knowledge of wild animals—a higher percentage than for any other wildlife and nature program. This really makes us feel that our efforts have been worthwhile.

One day in the spring of 1971 I received a letter from Chancellor H. W. Schooling of the University of Missouri that I was to be presented with an honorary doctor of science degree, the scientist's equivalent of a Ph.D. at their graduation ceremonies. As a college dropout from the University of Missouri, I felt that this was the most wonderful thing that could have happened to me at that time. I was to become *Dr.* Marlin Perkins. My father would have been gratified that I finally did get a degree, but he certainly never would have expected that one. That was followed by a second doctor of science from Rockhurst College in Kansas City, and then by a third from Northland College in Ashland, Wisconsin. In 1982 MacMurray College in Jacksonville, Illinois, conferred upon me the degree of doctor of public service.

While we were in southern Utah filming a story about expanding the range of the desert bighorn sheep from the south to the north side of the Colorado River, a helicopter landed on our site. General Jimmy Doolittle got out with a couple of representatives from Mutual of Omaha. He took me a little to one side and said that as chairman of the Mutual of Omaha Criss Award committee, it was his honor to inform me that I had been unanimously selected to be the recipient of the Criss Award for meritorious service. Dr. Benjamin Elijah Mays, past president of Morehouse College in Atlanta, Georgia, was to share the award with me. It was to be presented in October in Omaha. General Doolittle himself had been a recipient, as had Helen Hayes, Marian Anderson, Mr. and Mrs. Bob Hope, Dr. Howard Rusk, Dr. Jonas Salk, Dr. Thomas A. Dooley, William Gargan, W. Earl Hall, and Drs. Phillip S. Hench and Edward C. Kendall.

When October arrived, so did our children and many dear friends, who came to be with Carol and me on this very special occasion in my life. The ceremony was held in the fabulous underground, glass-domed addition to the Mutual of Omaha central headquarters. The Symphony Orchestra of Omaha was there to play, and the Boy's Town choir sang "I Talk to the Animals." The huge space under the dome was filled with flowers, tables and chairs, and people. Some of the artifacts I had collected from my travels throughout the world were exhibited in a museum section. Carol and I were seated on a dais with the chairman of Mutual of Omaha, V. J. Skutt, and his lovely wife, Angela. Twenty years before he had set this all in motion with his idea for Mutual of Omaha to sponsor "Wild Kingdom." With us were Dr. Mays, General and Mrs. Doolittle, Dolores Hope, and Irene Dunne.

General Doolittle introduced me and Dr. Mays, and then Dolores Hope, a Mutual of Omaha board member, presented the award. It was an evening of great honor for me. In my mind I stood back from all the splendor and wondered if it could be a dream. I remembered that little Perkins boy trudging home from the fields with a snake or a toad in his pocket—suddenly I remembered my mother and wished that she could know how my life had turned out.

In 1978 I received a call from Ron Blakely, director of the Sedgwick County, Kansas, Zoological Gardens. Ron had started in the zoo field after his graduation from Michigan State University, when I hired him as a zoologist on the staff of the Lincoln Park Zoo, and later he became curator of birds.

Ron was calling to inquire whether I would be attending the annual conference of the American Association of Zoological Parks and Aquariums, to be held in Denver that year. When I told him I was planning to be there, he mentioned that something special was going to happen at the banquet on the last evening of the meetings, and I was scheduled to give a short talk. If my plans changed, he asked me to let him know.

When that evening arrived, I was seated between Carol and Ron at the head table. Ron opened his remarks by saying that he and a number of others present had worked under me at the three zoos I had headed: Buffalo, Lincoln Park, and Saint Louis. He talked about my system of developing zoo staff, based on the idea that a curator or zoologist should not only receive zoo experience in his specialty but should also be interested in and understand other phases of zoo operation. Ron then spoke of the number of people in important jobs in zoos around the country who had received their training under me, and asked them to stand up. Twenty-two members of that audience rose to their feet. There were several others who could not be there in person but were with us in spirit. Then Ron read a special award citation:

THE AMERICAN ASSOCIATION OF
ZOOLOGICAL PARKS AND AQUARIUMS
PRESENTS THIS
SPECIAL AWARD TO
MARLIN PERKINS

In recognition of his meritorious service to the wildlife of this earth and to the zoocultural professions; by educating the public to the wonders of wild things and wild places through the presentation of their true stories; by putting to rest the myths and misunderstandings which hamper an accurate awareness of the wonders of nature;

by fostering a conservation ethic and by presenting wildlife appreciation and stewardship as worthy and noble precepts; and by depicting, through his own example, the zooculturist in a learned, dignified, and exemplary fashion, thereby ennobling the profession.

September 21, 1978
Denver, Colorado

Following this reading, Ron announced that the Board of the AAZPA had created a Marlin Perkins Award, to be presented to a member of the AAZPA who had made a special, outstanding contribution to the zoo profession. I looked out at that group of splendid young men, who were, and always would be, my boys, standing together smiling up at me, and I don't believe I ever had a prouder moment in my life. I glanced at Carol and she was wiping some tears from her eyes, and I must admit that my own emotions were almost overwhelming.

22

Elephant Adventure

Don Meier and his able longtime assistant Dick Reinauer had arranged for us to film a story about how wild elephants are captured and trained to work and about their labors in the forest. Warren Garst had set up the original concept with officials at Mumadulai, eighty miles from Mysore in the Nilgiri Forest in India. I had long wanted to film the story, and this was my chance. Cars would be rented in Bangalore; food and rooms were available in a rest house at the forest headquarters.

I checked out my personal still cameras, lenses, and film; packed light clothing for the hot weather; had my shots or boosters to prevent cholera, typhus, typhoid, and tetanus, as well as a shot of gamma globulin and a dose of oral polio vaccine. My smallpox vaccination was still good and so was the one for yellow fever. My stock of vitamins and trace minerals was replenished, and I started my weekly 500 milligrams of chloroquinine diphosphate three weeks before departure, setting Sunday as the easiest day to remember. Malaria was likely in India.

After a long flight that took me from Saint Louis to London to

Bombay to Bangalore, and a good night's sleep at each of these stopping places, I joined Warren Garst, Tom Allen, and George Bokland for the last leg of the journey to the place of the hunt, or *kedah*. In Bangalore we rented two cars, one to carry the photographic equipment and the other for ourselves. Leaving the traffic and business buildings of this thriving city, we were soon on a narrow, tree-lined tarmac road. Fields of vegetables and grain stretched on each side; people were working with primitive tools, some plowing with a wooden plow pulled by a bullock. Farm produce was being transported on the road in bullock carts or camel carts; a camel loaded with cut grass looked like a moving haystack until we got close enough to see the head and neck being controlled by a man with a rope attached through the nose of the camel. I was glad to have a driver who was used to maneuvering along these roads, coping with the slow speed of the carts and foot traffic sandwiched between the faster-moving bicycles, cars, trucks, and buses. We frequently had to pull off the pavement to clear a wide truck or bus.

Because of our late start we were obliged to spend the night in a lovely old colonial hotel in Mysore. The second-floor rooms, reached from an outside covered balcony, were huge, with ceilings at least fifteen feet high from which hung fans to circulate the air. The windows had curtains, draperies, and wooden shutters which could be adjusted to bar the sun's heat from the room but admit a breeze.

The dining room likewise was large and high-ceilinged, its walls decorated with large gold-framed oil paintings of several generations of the maharajas of Mysore. Royal-red draperies hung at appropriate positions around this attractive room, and ceiling fans cooled the air at the table level. The waiters were richly dressed in red and gold turbans, red jackets embossed with gold braid, and white trousers, tight fitting as is customary in India. The menu was varied; the food good. We enjoyed dinner.

Next day we paid a short visit to the zoo, privately owned by the maharaja and not far from his palace. The director took us around the grounds, and I was pleased to see Indian animals not commonly displayed in Western zoos, such as the Nilgiri tahr goat and the Ganges gavial, both endangered species. Much of the zoo was modernized, with moated exhibits; particularly well presented was an African section displaying giraffes, zebras, and various antelopes.

We could not linger because we still had a long drive on the crowded road to the Mumadulai forest preserve. When we got near we left the main route and drove through the entrance gates to the forest preserve headquarters. Soon after we had settled into our small but adequate quarters, the senior forest ranger in charge took us on a tour of the forest preserve. It was a rolling hilly country, and the thick forests had some open glades where we saw herds of the beautiful chital deer. The reddish-brown body of this handsome animal has whitish spots scattered close together over the back and side. Most deer are spotted as fawns, but the chital retains its spots through life. These deer were one of the prey animals of the five tigers that lived in the forest.

One road climbed to the top of a hill, which was bare of trees. There the working elephants were kept. A small building housed the facilities for storing and preparing the special concentrated foodballs, fortified with vitamins and trace minerals and made palatable to the elephants with a binder of molasses to hold the ground grains together.

Near this building was a row of heavy iron rings attached to a strong cement foundation sunk below the surface of the ground. The chain on the right front foot of each elephant was attached to a ring each evening when the animal returned from work. The rings were spaced out to allow each elephant room to move a step or two while feeding on the 300 to 400 pounds of tree trimmings and hay that each consumed daily in addition to the elephant cakes.

Each of the twelve elephants had his own mahout, a hill tribesman of the region. Their village was nearby, and the men lived there with their families. They spoke their own hill-tribe language and taught about sixty words of it to their elephants. The word *zeet* was the command to lie down. This command was to have a special meaning for me later on. Other words signaled rise, trunk up, move forward, move backward, lift the leg, raise your right or left front foot, turn right or left, pick up the stick, and so on. To mount his elephant a mahout asked the animal to raise a right or left front foot, then placed his bare foot just above the toenails of the elephant's foot; the elephant then raised its leg higher and the mahout walked up the elephant's leg, hanging onto the top of the ear for balance, until he was standing on top of the head; then he would sit astride the neck. (Sahibs are allowed to mount in a slightly easier fashion, with the elephant lying down.) When the mahout was in riding position, each bare foot was under an ear of the elephant and was pressed on the animal to guide it. The mahout carried a flexible stick five feet long and half an inch thick to discipline the elephant by striking it a sharp blow lengthwise at the base of the trunk between the eyes. The blow can't hurt much, but is an unpleasant reminder to the elephant that it has done something wrong.

The senior ranger drove us to another section of the forest where we saw a group of wild elephants from which two or three would be captured during the hunt we planned to film. They were feeding in a deep cleft between two hills; a small stream ran through the bottom. As we drove away we climbed to the ridge of another hill, and on the slope of the hill opposite we saw a small group of wild gaur. Farther on we were shown the elephant-catching area. A steep-sided ditch had been dug about four feet deep and six feet wide in a circle. A jungle path came by one side, and heavy logs had been laid across the ditch at that point to give access to the island in the middle. Freshly cut sugarcane would be placed in the middle of the circle day after day to lure the wild elephants there, but no one would be around to disturb them. An employee could make observations from a platform in a tree some distance along the path.

In order to be prepared for the day we would ride in to film the actual capture, Warren, as associate producer, set up a scene on the jungle trail. Two big tuskers with mahouts astride were stationed just off the path to await the arrival of Tom and myself. With both cameras in position, one for the master scene and the other for closeups, we were given the signal that the cameras were running. I led off, as I was to pass in front of Tom's elephant (named Arjun) and mount my own massive tusker, Ayyappan. Just as I was in front of Tom's mount, my mahout gave the command, *"Zeet!"* for Ayyappan to lie down in the mounting position. I kept walking, thinking I would pass the animal before he started to obey.

I woke up on the ground a few feet from the path. My first thought was, "What happened?" Then I saw my left hand was bleeding, and I tasted blood. Raising my head further, I rolled over with help from Warren and Tom and sat there spitting out blood and trying to understand what had hit me. Tom said, "Ayyappan tossed you out of his way with a tusk." Then I remembered I had been walking in front of my elephant, who had just been commanded to lie down.

In order to lie down an elephant must first step forward so that his forefeet are in front of him. At that moment, Ayyappan had simply reached forward and with a tusk flicked me out of his way. It was not a deliberate attack; I was just in the wrong place at the wrong time. I certainly couldn't blame the elephant.

I washed out my mouth with water from a canteen and I heard Warren talking to the ranger about the availability of medical attention. I then discovered my teeth had gone through both of my lips and my nose hurt and was also bleeding.

I felt like getting up, and with a little help I did. I could stand all right, and nothing below my waist hurt when I took a few steps. They washed my face and bleeding left hand with water from the canteen. My fingers moved without pain and I could see that the back of my hand had only been scratched by the gravel when I hit the ground, for I had been holding the shoulder strap of my binoculars with that hand to keep it from sliding down from my left shoulder. I explored my body for other injuries, and though my face began to throb, I could find no other damage.

"It's just near here. I'll put him in my Land Rover and we can be there in fifteen minutes," I heard the ranger tell Warren. We drove to our destination, a sizable hospital building. A long series of steps led upward to the entrance. Warren said, "Marlin, do you think you can walk that far?"

"I'll try, but only one at a time, please," I replied.

I started off slowly, with Tom and Warren each holding an arm. About halfway up I felt a pain on the right side of my thorax, low down on the back side. But this pain increased only slightly as I continued up the steps and into the hospital.

Two English-speaking Indian physicians met me in the emergency room and were told by Warren and Tom what had happened. When they moved my nose gently, both the doctors and I knew it was broken; I could feel the bones of the break moving against each other. The examining doctor felt my cheeks; they were both sore. Next he gently applied some black tarlike material to all the abrasions. Long narrow rolls of gauze soaked in the same disinfectant were inserted in each nostril. The surface injuries were then dressed, and my face and hand were finished for the time being. It was then that I told the doctors of the pain in my ribs. Examination revealed three broken ribs. The doctors explained there was nothing to be done, as the ribs would heal of their own accord. I must just be careful not to move them any more than I had to. The doctors asked me to return the next day to have the wounds dressed again.

When I got to the rest house where we were staying, I talked with Warren about the filming schedule, as obviously I could not appear on camera for some time. My ribs hurt more than my nose and lips by this time, and I lay down on my bed very slowly, on my left side. I found I could be comfortable also on my stomach with my head turned to one side and without a pillow. I took another of the pain pills the doctor had given me before trying to sleep.

Next morning I took another pain pill, as I hurt all over and had a headache. After a liquid breakfast taken through a straw, also supplied at the hospital, I had another talk with Warren. I was worried about the three broken ribs and the possibility of further injury to the tissues around the jagged edges of the breaks. I remembered that there was an American doctor teaching at the university in Bangalore, and suggested that I go in by car and have him check my injuries too. Warren agreed, and after a visit to the hospital and new dressings, I was ready to travel. Warren would continue to film with Tom, and I would return as soon as my face healed.

The American doctor checked me all over. He admired the gauze nose plugs and indicated that the black disinfectant, while no longer widely used in the United States, was highly effective. My wounds were not yet infected, nor did he think they would be.

He verified that three ribs were broken, but as far as he could tell without an X-ray, they were in good position. He added that even if I had X-rays taken there was little to be done short of surgically opening my thorax and positioning and securing the ribs in exact alignment. He thought they would heal well without this. So back to my comfortable hotel, where I could rest and have my meals sent to my room.

I was surprised at how rapidly the damage healed. At the end of five days my nose and lips looked nearly normal; I had quit taking pain pills and felt pretty good. So back I went to Mumadulai to finish the elephant *kedah*.

On the sixth day after the elephant Ayyappan had tossed me out of his way with a tusk so he could comply with his mahout's command to lie down, I mounted him and finished the scene we had started. When I had to pull myself up with the rope around his chest, I was painfully aware of my three broken ribs.

Two days later the capture took place. Two wild elephants had been confined to the island in the middle of the moat by having a tame elephant walk up and remove the log bridge onto the outside path. When all elephants and personnel needed for the capture were on hand, the bridge was reinstalled. One at a time two tame elephants crossed and were maneuvered along both sides of a wild animal, pressing closely upon the wild one. A rope the size of a ship's hawser was dropped over the wild elephant's head and trunk and secured around its neck. The other end was spliced into two ropes so that each branch could be secured around the necks of the two tame elephants. Another rope was attached to the wild elephant's hind leg by a man on foot and to a third elephant being controlled by his mahout.

In the days that followed, several other wild elephants were captured in the same way. Then the captured elephants were led back and secured to trees in the same area where the tame elephants were kept. It would not be long before they, too, were accustomed to working for humans and could be used to entrap their wild relatives in the jungle.

Carol had not come along on this expedition. The day before I left for India she had had a small growth removed from her face, as an outpatient, by a plastic surgeon. She was feeling fine and there was no concern about my leaving. After a few days the biopsy report came back and she learned she had malignant melanoma, an extremely virulent form of cancer.

She made sure no one would tell me and alarm me and went off alone to the Mayo Clinic. Our daughter Marguerite was at home. I called Carol from India while I was recovering from my accident with the elephant, but I didn't tell her about my problems either. We both told each other we were fine.

Imagine our surprise when I got home and Carol and I both had scars on our faces! Carol had had quite a bit of surgery on her face, but thankfully she was one of the lucky ones. Having caught the malignancy very early, she could be hopeful of a full recovery, and so it has turned out.

We had to laugh at our mutual deceptions so as not to worry one another when we were so far apart—and we could laugh because all's well that ends well.

23

At Sea with the Humpback Whales

At the end of six years, Jim Fowler decided to leave the show to carry on other work. I was happy when later on, at the start of the 1982–1983 season, Jim decided to return to our program as co-host.

Stan Brock, who took Jim's place in 1968, had been manager of the vast Dadanowa Ranch in British Guiana, renamed Guyana when the region became independent. With his experience as a barefoot cowboy and some knowledge of the wild animals on the ranch, he became my assistant and co-host of "Mutual of Omaha's Wild Kingdom." Born and raised in northern England, Stan's accent was a notable change for the show, and he was admired for his athletic physique and great strength.

Four years later, when Stan departed from the show, he was replaced by one of Ross Allen's sons, Tom. Tom Allen had been a frogman in the navy, so he was an expert scuba diver. He also had learned much about animals from his father at Silver Springs, Florida.

Tom grew up working with and handling snakes, lizards, turtles, alligators, and many other animals on display at his father's Reptile Institute. He was just what "Wild Kingdom" needed at that time. One of our photographers, Ralph Nelson, had worked in a California shop that sold diving gear. He not only knew much about diving and marine life, but had designed and built an underwater housing for an Arriflex camera with a 400-foot magazine. With Ralph, we had filmed several underwater shows, and now, exploiting Tom's knowledge and ability, we expanded that phase of "Wild Kingdom."

One such show was filmed off Islamorada Key in Florida. We were staying at a motel on the main highway and awoke one morning to find ourselves beached by small-boat warnings due to a high wind. Tom suggested we take advantage of the enforced inactivity to practice some diving techniques in the motel pool. This pool was ten feet deep and large enough so that we could get additional underwater experience in it. Tom set up a scuba class for all of us, including Don Meier, who since he had no scuba training, had up to now stayed in the boat when the rest of us dove. Learning to dive would enable him to be with us underwater, understand the filming segments, and better direct the story line.

Tom started us with the basics and worked us up to more difficult activities. For example, he put my tank, weight belt, and flippers on the bottom in ten feet of water. My job was to get them on under water. I was allowed my mask when I dove from the surface to the bottom. With one hand I turned on the air valve, put my regulator in my mouth, cleared it of water and started to breathe; with my other hand I hung onto my weight belt to keep me from rising to the surface. With some effort I was able to place the belt around my waist and secure it there with the buckle. Then my arms went through the loops of the backpack fitted to the tank, and I secured the waist belt with another buckle. This was not as easy as it sounds, for the tank was free to roll and so was I. The last thing to put on were the flippers. Just after I had accomplished that and had righted myself into the proper prone position for swimming, Tom appeared in front of me and signaled (by running the finger end of his flat hand across his throat) that he was out of air. I took a deep breath, held my breath, and handed Tom my mouthpiece. He took a couple of breaths of my tank and handed the mouthpiece back. I remembered to clear it of water by blowing into it with my own air and started to breathe. We continued to buddy breathe for a while and then Tom gave the signal to surface. This and other lessons made me ever more comfortable in the water.

While we were filming underwater near a large sea-bottom habitat used by scientists off Grand Bahama Island, a scene was set up that showed Tom and me approaching in a small submersible in which we sat, breathing from our scuba air tanks attached to the open metal frame. Ralph Nelson and Rod Allin dove ahead of us and were sitting on the bottom with their cameras. When we stopped on the bottom fifty feet below the surface, I slid out of my seat, took a good breath of air, and leaving my tank in the sub, turned and swam to the big underwater habitat thirty-five feet away. The manhole through which to enter it was toward the back, with the opening toward the bottom. There was only about a foot of water above the manhole so it wasn't long before I surfaced into the air of the interior of the habitat.

Ralph, Rod, and Tom came to join me and one of the research scientists inside the habitat. In the course of our conversation Tom asked me if I had ever made a free ascent; that is, without an air tank. I hadn't, and they all agreed this would be a good chance for me to learn the technique in fifty feet of water, in case I ever really had to use it.

Tom reviewed the details. Tom would leave the habitat first, with his scuba tank on. I was to get on my mask and flippers and join him at the little submersible resting on the bottom. I would take a breath of air from my tank, which was still attached to the submersible, and Tom, wearing his tank, would face me and hold me by my arms near my shoulders. With our faces close together we would start our ascent, Tom watching me closely to make sure I continued to exhale slowly through my closed lips while I softly hummed. If I should exhale too fast and run out of air we could buddy breathe to the surface from Tom's tank. The rule is that you must not rise faster than your bubbles, to keep from getting the bends, and you must continue to exhale all the way to the top. I was afraid I would run out of air before I reached the surface, but Tom reminded me that at fifty feet below the surface the air in my lungs was compressed, and as I rose in the water it would expand. If I held my breath without exhaling, the air would expand so much that it would become larger in volume than my lungs could hold, and my lungs would rupture.

With all this clearly in mind, I swam to my air tank, breathed for a short while, took a final big gulp of air, and turned to face Tom. Slowly we started upward as I began exhaling with a hum through my lips. Tom gave me a nod and a thumbs-up sign to let me know I was doing it correctly. My worries about running out of air vanished and I always had a small stream of air bubbles rising in front of my mask. Within about fifteen feet of the surface I looked up to see how close we were. I felt Tom touch me on my lips to let me know I had stopped exhaling. I started again, and still had air in my lungs when I broke the surface. I climbed into our waiting boat while Tom went back down to get the sub and Ralph and Rod. As I sat waiting for them, I mentally reviewed my free ascent and realized I had acquired a safety technique that I would remember the rest of my life. It was a good feeling.

This training was helpful when I joined Dr. Howard Winn in the Caribbean. For several years Dr. Winn, of the University of Rhode Island, had been studying humpback whales, and we were invited to film an episode with him. The studies were being conducted on a research boat outfitted with special equipment to record the eerie songs of the humpback whale. We were to scuba dive with the whales and film their activities both from underwater and from the deck of the boat.

When all the logistics for this episode of "Wild Kingdom" had been worked out, I flew to San José, Puerto Rico, and joined Ralph Nelson and Rod Allin. Together we boarded the research ship with all our gear and settled in for a long stay at sea. We headed for the Silver Navidad banks, a shallow area where the whales come to give birth to their calves. Here the water is warm compared with waters off the far northeastern coasts of North America where the whales go to feed during the summer months.

As soon as a whale calf is born its mother helps it to the surface, where it takes its first breath of air. Staying close to her, the calf soon begins nursing on its mother's rich milk, ejected by strong muscles directly into its mouth. At first the fifteen-foot-long calf has little fat on its body, and for this reason it needs the warmth of the Caribbean water. But it adds weight quickly because of the exceptionally high fat content of the milk. By the time the whales' northward migration to the feeding waters of the Arctic region begins, the calf has added enough fat to insulate it from the cold of the northern waters.

Not long after we reached the Silver Navidad banks the hydrophone on the deck of the research ship was receiving humpback whale calls and these were being recorded. Some of the whales, which average some fifty feet in length, were also within sight of the ship when they surfaced for air and slapped the water with their long white flippers. Many photographs of those activities were taken, including pictures of the tail flukes as the whales raised them high above water before sounding (diving). These photos were to be studied for irregularities or black-and-white markings that could be noted and used to identify individuals.

The research vessel had a diving platform at the stern. From this we could drop into the water ahead of the whales in hopes of getting close enough to film them under water. Sometimes the whales would turn in another direction, and we would have to climb back onto the platform and wait for another chance. Many times, though, we were successful, and on a few occasions the photographers were so close they could not see the whole whale in their viewfinders. On one of these occasions Ralph Nelson got so close that the narrow leading edge of a flipper hit his hand, cutting to the bone on one finger.

To try to keep up with the whales or to get to them before they moved away, we had battery-powered underwater divers' propulsion vehicles. Holding on with both hands, we were propelled forward faster than we could swim, varying the speed to suit the occasion. There was a motor noise, however, and this warned the whales of our presence and they would swim away. We had better luck by remaining on the surface breathing through our snorkels and waiting for whales to swim near. When they were close enough we dove from the surface, breathed the air from our tanks, swam down to the level of the whales, and waited for them to come near enough to be photographed.

One day we decided to try using a rubber boat with an outboard motor in conjunction with the slower mother ship to see if the fast rubber boat could speed ahead of the whales and turn them back toward the research vessel. We were relying on the noise of the outboard motor to cause the whales to turn. Several of us went over the side, joining three others in the water, and submerged, hoping to get good photos as the cetaceans turned.

I gazed ahead of me through the luminous blue water. Never in the wildest thoughts of my formative years had I imagined that one day I would be swimming with a pod of humpback whales. Yet here I was, in scuba gear, watching a group approaching, a mother and calf in the lead. I remembered an article by Jon Lindbergh in which he declared that just to have this experience was one of the goals of his life. I knew Jon would have enjoyed, perhaps been thrilled, to see the mother and young whale come slowly toward us and then veer off to her left to join the rest of her pod. How effortlessly she and her calf moved through the sea, mysteriously propelled. The long, thin, white flippers stood out in sharp contrast to her dark body. The ridges of the lower jaw were visible. They were the pleats that allowed for the expansion of the skin between her fifteen-foot mandibles when she opened her mouth like an enormous cavern as she swam through a sea in the far North teeming with krill and small fish. The only visible indication of that enormous capacity was the line that marked the extent of her mouth. Her calf was staying close by. From a position below its mother it rose to near her head and then shifted to just above her body. As the pair moved off I could see the slow rhythm of the tail flukes as they sculled up and down, propelling the whales forward away from us to become obscure in the hazy distance.

On board the research ship, Rod Allin was to operate a camera to film whale activity. But the captain misjudged, communications failed, and as the whales moved along, so did the ship. The six of us underwater heard the sound of the ship growing fainter, so we surfaced. We were aghast to see the ship rapidly moving away from us. We inflated our vests and floated in a group watching our rescue ship growing smaller and smaller. We waved frantically, trying to attract attention, but no one was watching. We realized that the ship might get so far away that we would be invisible, bobbing up and down between the waves. We were more noticeable when we stayed together, but six men in a trough between waves are not highly visible. And that could be the moment when someone looking in our direction would not see us. Thoughts ran through our minds that we might be there a long, long time. When our shipmates realized our plight and tried to retrieve us, they would not know exactly where we were and might cruise the ocean for miles around and not find us. Although we did not panic, our hopes were sinking as we realized the seriousness of our situation.

Then as all six pairs of eyes watched the ship, it made a 180-degree turnabout and headed in our direction. Rod Allin in his elevated position had turned to look for us; we were luckily on a crest of a wave and he saw us. He called immediately to the captain in the wheelhouse below him and kept his eyes on us so he would not lose the spot. The captain turned the ship and Rod continued to direct the captain until he, too, could see us.

This, I think, was the most dangerous situation I have ever been in in the Wild Kingdom.

24

Africa Once More: Smugglers and Wild Chimps

In September 1976 Carol and I flew to Tanzania. When we arrived at the airport of Dar es Salaam and customs had been cleared, we took an old, beat-up, tired, carbon-monoxide-breathing taxi into town, where we registered at the Hotel Kilimanjaro. We were missing one suitcase, mine—the one that contained the drip-dry suntans, boots, and other things I needed for the filming of "Wild Kingdom." A tracer had been sent and we hoped the case could be located and delivered to us before we had to leave the following day for Kigoma.

In our room I measured with my eye the light-tan pants suit Carol was wearing. As she turned, she read my look and said, "Wait just a minute. Maybe it *will* fit you, and if it does, you're welcome to use it for the film." (It's amazing how well many wives know their husbands.) The pants suit was a little tight in a couple of places, but it was a knit jersey fabric I could get into, and it was more suitable than my regular street clothes even if it was designed for a female figure.

That problem had been solved, but we had to go to the markets to try to find some supplies for the camp at the Mahale Mountain Chimpanzee Reserve. A note awaiting our arrival from Warren and Genny Garst had told us of their need for food—meat, bread, rice, but especially salt—and had asked us to bring what we could. The only meat we could buy was a salted ham with the hoof still intact. Carol bought it and had the hoof cut off. We managed to locate some salt, but only a small amount. In one store we found a few cans of corned beef, canned beans, and in another some milk powder. It wasn't supermarket shopping, but we did the best we could.

The following morning we taxied back to the airport and checked in for the once a week cross-country flight to Kigoma on the shore of Lake Tanganyika. We were in the airport building waiting for our flight to be called when a plane landed and taxied on the apron near the waiting room. We watched the passengers walk down the steps and across the apron to the customs entrance. Then the carts piled high with luggage were wheeled to the building. Without much hope I watched for my bag. And there it was—on the very top of the pile. I rushed to the baggage claim area, contacted a customs official, and retrieved my bag in time to check it on our flight to Kigoma. What a relief. I wouldn't have to wear Carol's pants suit after all.

We flew over much of the same route that Stanley had traveled when he was looking for Livingstone, except that he was walking, with a great long line of porters, and it took him months to complete that part of his journey. With three stops, our flight took four hours. Carol, with her keen interest in Livingstone, had looked forward to flying over the forested areas Livingstone had seen a hundred years ago. Instead, much was semidesert, for the forests had been cut down for their wood and to make fields for agriculture. It was grim evidence of the problem facing the emerging countries as their population increases and creates an ever-expanding need for more food.

When we landed at Kigoma and walked to the small airport building, we were pleased to see Jane Goodall, her husband Derek Brysen, and son Grublin, who were returning to Dar es Salaam on the plane on which we had arrived. The previous year we had planned to film with Jane and the chimpanzees she had been studying north of Kigoma at Gombe Stream Game Reserve; then terrorists came ashore at her camp, kidnapped several students, and held them for ransom. The feeling was that that area was still dangerous, so instead we were traveling south on the lake to the Mahale Mountain Chimpanzee Reserve where Jane's colleague, Dr. Junichiro Itani, a Japanese primatologist from the laboratory of physical anthropology at Kyoto University, had been studying chimps for several years.

An American Methodist missionary took charge of us and suggested a nice clean rest house at their mission instead of the rundown Kigoma Hotel. He took us and our luggage there, and we were grateful, for when we went shopping in town for bananas and vegetables, we passed the hotel and saw the tremendous contrast. We were also invited for lunch with the missionaries and learned they would soon close their mission in Tanzania and leave the country.

We made contact with Dr. Itani, who arranged for us to leave in late afternoon for an eighteen-hour boat ride down the lake to his research station, where we would join Warren and Genny Garst. Dr. Itani would be busy most of the day in Kigoma getting the boat ready to sail.

We were to travel overnight, as Lake Tanganyika is the second deepest lake in the world (the deepest is Lake Baikal in Russia) and during daylight hours there are high waves on this 450-mile-long expanse of water. It is calmer at night when the winds have died down.

We tried again to buy supplies for camp, but could not find salt. Our new friends explained that the principal source of salt for the country was a mine only forty miles north of Kigoma; but no salt was available in Kigoma because it was shipped across the lake to bordering countries where it was sold on the black market. Thus the socialist ideology was breaking down as individuals lined their pockets.

We could see the harbor from our rest house. Carol kept a lookout for the ship that would take us to the Mahale Mountains. She knew about a beautiful steamer on Lake Victoria, with staterooms, bathrooms, and deck space for walking and viewing the sights as the ship moved along. She also had seen the film featuring Katharine Hepburn and the *African Queen*, a sizable boat on which there was surely some place where a lady could retire in privacy. She was not prepared, therefore, when our Land Rover stopped about two blocks away from the edge of the lake and our luggage was carried over the rocks up a little incline. Following, we got our first look at *our* ship. Carol stopped dead in her tracks with an unbelieving expression on her face. Finally she said, "Oh no! that can't be *our* boat, can it?" Almost afraid to answer, I could only murmur, "Well, let's go and see." As we walked forward, the question was answered for us, as our luggage was being put aboard. When we got closer, there too was Dr. Itani with one of his students. He welcomed us cheerfully and suggested we would be ready to depart in a few minutes.

This boat was a larger version of a rowboat, but had a corrugated-tin roof supported by short posts. We could see the outboard motor and a couple of drums of gasoline resting on the bottom of the boat. Along each side were shelves about two feet wide. These made it possible to crawl from the stern to the bow, about eighteen feet away. These were also our bunks. We were to sleep on them and sit on them for eighteen hours.

After a quick glance at our accommodations we both decided we should take a walk behind the bushes back a little way from the shore.

When we returned we waded out to our boat and climbed back on board. We crawled forward on our hands and knees. Dr. Itani gave us each a somewhat soiled blanket. Fortunately, we always travel with our own down pillows. Carol unzipped the canvas bag and brought them out for our long night.

The Yamaha outboard came alive; we started moving and soon lost sight of the buildings of Kigoma. Our route lay somewhat offshore as the three Africans steered the boat in a straight line for our destination. As dusk settled we lost sight of all land; the waves gave a gentle rolling action that kept us moving our muscles to counteract the tendency to roll either over the gunwales into the lake or onto the pile of freight being supported by the ribs of the bottom of the boat.

We had had an early dinner with the missionaries before departure and they had given us a loaf of homemade bread and some fruit, which we enjoyed about nine o'clock.

Dr. Itani offered a swig of locally made gin from his bottle, which we declined. He and his student, however, had several drinks, refusing our offer of bread and fruit; then they rolled up in their blankets and were soon sound asleep.

We took stock of the situation. Dr. Itani and the student were lulled into a peaceful sleep. The three tribesmen from the Mahale Mountains, smoking and nipping from their own bottle of gin, were conversing over the roar of the engine. One was steering. Assessing the situation and taking no chances, Carol downed a Dramamine and a sleeping pill, rolled up in her blanket, and lay down. I tried to stay awake, but after a while I too rolled up in my blanket and laid my head on my soft pillow.

I'm not sure how long I slept, but I was suddenly wide awake from the strong smell of gasoline. Looking around I saw the glow of three cigarettes at the back of the boat. I switched on my flashlight and quickly saw that the bottom of the boat was awash with several inches of gasoline. I called to Dr. Itani and when he stirred awake explained the situation.

"Tell the men to put out their cigarettes and bail out the gas," I shouted, showing him the gas with my flashlight at the same time.

He immediately called to them in their own language; they came forward and found a leak in a gas drum which was upside down in the boat. They righted the drum, bailed out the gas, and casually lit up cigarettes as we pushed along again. Carol woke up briefly and smelled the gasoline, but the effect of her sleeping pill was strong, and she was soon breathing deeply again. I looked around with my light for life preservers. There were none. I then looked for something that would float. No loose boards. Nothing I could see that would support us if we had to abandon the boat. I spent an uncomfortable hour trying to figure how we would stay afloat and try to reach a shore of the lake, with no land in sight.

I must have dozed off, for suddenly I was awakened again as our motor stopped. The Africans borrowed my flashlight and went to work on the Yamaha. Thirty minutes later it roared into life and we were underway again.

Around midnight we awakened once more as the motor changed pitch; we realized we were approaching a shoreline. The Africans cut the motor; two of the tribesmen stepped onto the lake bottom and indicated by sign language that we were to do the same. It was our rest stop. I stepped down first in water almost to my knees and relayed to Carol the slipperiness of the round rocks underfoot. With a couple of helping hands she eased herself off the stern of the boat, but it was a little deeper than she thought. Her foot slipped, and she sat down in the water. Carefully we waded ashore, and against the starlit sky we could see the outline of some bushes. Carol started to move behind them when she squeezed my hand hard and whispered in my ear, "There is someone lying there, I can hear him snore."

Standing still and listening, we heard others nearby. We had come ashore to a beach used by smugglers. So back we went to the boat, and as we did we could make out another boat nearby. This ended our hopes for a rest stop.

When daylight arrived we saw two boats like ours on the lake a quarter mile and a half mile from us. We were all cruising at about the same speed. We watched them and they watched us. It was a little eerie, as we remembered the recent kidnapping at Jane Goodall's camp at Gombe. The two boats contained smugglers and they were certainly keeping a sharp eye on us.

After a time we approached some hills, and farther in the distance were mountains, all hazy and gray. I scanned the shore with binoculars. When we were opposite the mountains we began to see platforms of leaves and branches in the trees. Chimpanzee nests. Chimps make new nests each night, and as we progressed we saw dozens of them.

About noon, we rounded a point, and there on shore were some tents, and Warren and Genny waving at us. As we landed Warren said, "Hi! Come and meet the welcoming committee."

We walked past the tents, and a short distance down a path we saw five chimpanzees standing on their hindlegs to get a better view of us above the high brown grass. They were the first chimpanzees Carol and I had ever seen in the Wild Kingdom. What handsome animals!

"Let's walk along the path toward them," Warren said. "It will lead us to the chimp picnic area."

The chimpanzees preceded us as though they were indeed showing us around. We came to two sheds, and there under the trees were more chimpanzees and Dr. Itani's caretaker feeding them short sections of sugarcane. The animals were all sizes—one large male, several mothers with babies, youngsters playing together, and still others we could see and hear in the trees nearby. What a welcome this was for us!

Our gear was brought ashore and placed in our tent, which had palm branches covering the roof to keep it from getting too hot. In a corner was our own private bath—a pan of water. We toured the rest of the camp, including the octagonal metal building where Warren kept his photographic gear. This was located under some trees, which shaded it. Down a path a sign pointed to the outdoor rest room. The facilities seemed to us luxurious, as they were enclosed in a small tent.

A quick lunch was prepared as we sat in canvas chairs and learned from Warren about the filming that had taken place and of the plans ahead. Dr. Itani and his staff had cooperated fully, and we were certain to see and hear wild chimpanzees every day.

Next morning we all climbed a steep, winding path through the forest. At switchbacks, as we paused to catch our breath, we could look down on camp and the wide expanse of Lake Tanganyika.

Smaller mountains bordered the lake shore, and we could see many chimpanzee nests as we looked down on the forest.

Near the top of the path Dr. Itani and his caretaker started to call to the chimps in their own *whoo, whoo* language. An answering call came from far around the mountain. We reached an open area blocked on one side by a large tree trunk blown down years ago. This was another feeding ground, and as it sloped in the opposite direction we could see the other side of the forested mountain falling away to a valley beyond which rose another mountain.

The calls of the wild chimps grew louder as the animals moved through the forest toward us. Dr. Itani and the caretaker continued calling to encourage them.

Warren had set up his camera and asked us to sit on a log with the food. He checked the light and the composition of the scene and waited. Carol stood by with a still camera. Genny had set up the tape recorder and was capturing the calls of the chimps as they moved toward us. The rustling of some leaves on an obscure path drew our eyes to the base of a large tree, and then we saw the form of a large female chimp walk past the tree and pause to look around and assess the situation. She must have recognized everyone except Carol and me, but as we remained still and offered no threat, she sat down. As she did we could see a baby clinging to the hair of its mother's stomach. As the baby chimp was now in an upright position, it too looked us over, then turned back and started to nurse.

Soon another, and then another, chimp came out of the bushes and took up similar positions. More followed until there were fourteen animals in the clearing with us. They approached the log and were handed pieces of sugarcane. Some came from behind the log and some walked along the log to get the food. They accepted Carol and me as they had Warren and Genny, and even the young animals took food from our hands. The feeding periods gave Dr. Itani and his students a chance to get to know each individual animal and eventually to sort out their relationships to each other. Each chimp had a name, and each could be recognized by some set of physical characteristics—a droopy lower lip, dark marks on the face, a torn ear, lighter hair on the back of an older chimp.

Suddenly a low alarm grunt, and all the chimps took to the trees. "Shibongo is coming," said Dr. Itani. "Big Daddy." Shibongo was male leader of this troop and the father of most of the babies. All the animals respected him and deferred to him, allowing him to eat first and heeding his moods. Dramatically as a king entering his audience chamber, Shibongo parted two branches of a bush, paused to look around at his subjects, and, head held high, strode forward on all four legs with measured tread. What a magnificent specimen—big, imposing, self-assured. He had the bearing of the one in charge, the Big Daddy. He stopped, looked around at the chimps in the trees, and then, satisfied that all was well, walked straight to the caretaker, accepted a piece of sugarcane, sat down, and ate it. Dr. Itani then offered him another section, which he took and ate. When finished, he walked past us, jumped on the log near me, and accepted food from my hand. As though on cue the other chimps came down from the trees. The female who had been the first to appear walked slowly over to Shibongo, smacked her lips in friendship, and offered to groom him. As though he had expected it, Shibongo closed his eyes. With smacking lips and many rapid soft guttural movements of air in her throat, she parted the hair of his head and face, looking for particles of food or skin which she removed with her extended lips. If she should find a tick or a flea, she would remove that too, but normally chimpanzees and other primates do not have fleas. As she moved to groom other parts of his body, Shibongo would raise an arm or roll over or turn around.

Other chimps came up to be near and watch and take part in the grooming of Shibongo. Eventually the mood changed, and as Big Daddy moved away from the circle, nearly grown to medium-sized male chimps edged further away from him, while keeping close watch on the direction in which he was moving. Immature animals of both sexes were playing games in the high branches of the trees. One would run out on a branch toward another in fake attack. If the one being attacked responded boldly, the attacker would turn and run or drop down to a lower branch. Like puppies, one would chase a companion who, allowing himself to be caught, would join the attacker in a mock battle of pulling and twisting with mild bites and open-mouthed laughter as one would tickle the other. It was a happy play time in the finest jungle gym in the world.

On another day Warren filmed Dr. Itani and me observing a chimpanzee approaching a tree that harbored a termite nest. A knot in the trunk of the sloping tree had a small hole in it. The chimp broke off a small twig, removed the leaves and bark with his teeth, inserted the small end into the hole, and waited with the intense interest of a fisherman. Then he slowly withdrew the twig and ran it through his lips, taking off the termites that clung to it; we could see his jaws move in eating. Then, again holding the twig carefully between thumb and finger, he inserted it once more in the hole. Over and over the chimp thrust in the twig, pulled it out, and ate the termites clinging to it. This behavior was first observed and recorded by Jane Goodall at her Gombe Stream camp. She saw a chimp tear the edges from a wide blade of sword grass, insert it into a hole in a termite mound, pull it out, and eat the termites that clung to it.

In the early literature about animals the distinction often used to separate man from the "lower" animals was the mistaken notion that man was the only creature that used tools. Animals have been more carefully studied since the days of the early natural histories, and we now know that a number of animals use tools. A marine otter brings a flat rock to the surface, rolls over on his back, places the rock on his chest, and breaks open sea urchins and shellfish by hitting them on the rock to get the flesh inside. A banded mongoose breaks an ostrich egg by throwing rocks at it with his hindfeet. The Egyptian vulture breaks eggs with a rock held in his beak. A finch in the Galápagos Islands uses a cactus spine held in his beak to probe for and extract insect grubs from holes in trees and shrubs. A chimpanzee not only fishes for termites with a twig or a blade of grass, but throws sticks and rocks as weapons and uses a stick or pole as we would a crowbar to exert leverage.

Our last morning in camp was spent in packing our gear. Toward noon, when this was completed, Dr. Itani invited us to his main camp, where he and his staff stayed. Warren filmed some transition scenes here, and after a leisurely lunch Warren, Genny, Carol, and I took some still pictures and then waded along the sandy-bottomed beach, always keeping an eye out for crocodiles.

Dr. Itani had received a gift package from his wife in Japan the day he came in the boat to Kigoma to get Carol and me. This contained Japanese foods and seasonings and a bottle of saki. He spent the afternoon preparing a Japanese meal. At the same time his staff of tribesmen killed a pet duck we had seen around camp. They stewed the duck and in another pot prepared mealy-mealy (cornmeal mush), their staple diet.

At dusk, when everything was ready, we sat down outdoors at a table and, starting with saki, ate course after course of delicious food. Dr. Itani was a gracious host as well as a fine cook.

After we had finished Dr. Itani informed us that we were expected to join the Africans and partake of their food. Not to do so would be taken as an affront by our hosts. So over we went and squatted down around the two pots and watched. Being Muslims, the Africans ate with their right hands only. Each person put his left hand behind his back, reached into the mealy-mealy with his right hand and rolled a ball the size of a walnut, dipped the ball into the duck stew and popped it into his mouth. I followed suit and found the stew a little too spicy with a kind of red pepper, but otherwise palatable. Carol, on the other hand, had been feeding the duck earlier in the week and vowed she would not eat it. In order not to insult our hosts, she went through all the motions of rolling a ball and dipping it slightly; then, figuring it was dark enough so no one would notice, she raised the food to her mouth and carefully dropped it inside the open neck of her shirt.

After we had expressed our thanks for the delicious meal, Carol disappeared into the shadow of a nearby tree, and scooping the food out with her hand (right one, of course) dropped it onto the ground. Before departing for our return voyage the next day, she washed her clothes in the lake. They were dry long before we reached Kigoma.

Another remarkable study of the social and community organization of primates is being carried on amid the ancient ruins at Polonnaruwa in Sri Lanka, formerly Ceylon, by Dr. Wolfgang Dittus of the Smithsonian Institution. To film Dr. Dittus's work with the toque macaque monkeys of Sri Lanka, the "Wild Kingdom" crew flew to Colombo and drove from there through the countryside of this jewel of an island, passing the ruins of ancient cities, crossing valleys green with rice fields, and wending our way past hills and mountains. I was accompanied on this trip by my nineteen-year-old grandson, Peter Brentlinger, who had developed a keen interest in photography. I knew that on this "Wild Kingdom" trip Peter would be able to develop his skill in photography under a variety of conditions. I have always felt it important to share our adventures with our family whenever possible. Peter and his brother Christopher, whose special interest is in music, and his mother, Suzanne, had been with us recently on a long safari in Africa.

Dr. Dittus had been studying the toque macaque monkeys for ten years. The toque macaques are indigenous to Ceylon, southern India, and parts of the East Indies and the Philippines. They get their name from the hairy topknots on their heads. Medium in size and brownish in color at Polonnaruwa, but sometimes gray elsewhere, these monkeys have very long tails. Dr. Dittus can recognize 500 individuals by sight. He knows the troop number that has been assigned to each individual, as well as its mother, its brothers and sisters, and its age. All the monkeys have been given names. In this locality there are eighteen troops, each consisting of from six to forty-five individuals. Dr. Dittus's system of spending enough time to learn to recognize individuals by sight seems to me far better and less degrading to the animal than the system of attaching color-coded ribbons to their ears or around their necks, used by some observers. The sign-recognition method is used also by Dr. Itani and by Dr. Ray Carpenter in his studies of red-faced macaque monkeys at Takisakiyama in Japan.

With my interest in archeology, the fact that the study area was in the ruins of an ancient walled city, about a mile and a half long and a half mile wide, built in the early twelfth century A.D., made the expedition all the more fascinating.

As we were driving into the area, Dittus stopped to look at a monkey. He raised his binoculars and after a moment said, "That is Lucy. She belongs to troop number six." I asked how he could be so sure from just a quick glance. In reply he opened his case and from a file for troop six extracted a card with Lucy's name on it. He showed me three print-out drawings of a monkey face—right profile, left profile, and full face.

"See here—the dark line on her lower lip?"

I looked at the card and then, through my binoculars, at the monkey. Sure enough, she had a dark lower lip. A dark spot on her left temple was also properly recorded on the card, as was a small tear in her right ear. Dittus then noticed another dark spot near her left eye, which he quickly marked onto the front-face view. The card contained other information about Lucy: Her mother was Helen, the dominant female of the troop, which gave Lucy a much better than average chance to become dominant herself. Lucy was about eighteen months old, and Dittus thought she would be mating in another few months. Lucy looked up, saw that the rest of her troop was moving off, and ran to catch up.

Soon we were filming the monkeys feeding on the yellow flowers of the cassia trees. In the background was the domed upper part of a stupa, called here a *dagaba*, with a cupola and tapering spire capping its summit. I found myself looking away from the monkeys at the perfect symmetry of the stupa and remembering others in Nepal, particularly the huge one at Bodhnath that can be seen for miles above the surrounding valley.

The monkeys spent hours each day searching for food. In an open grass meadow in a part of the old city they would rummage in the grass and on the ground for seeds and other edible parts of plants. We spent hours filming this activity in the meadow, through the scrub growth and trees, and at a water-lily pond, where the lilies and other water plants were eaten. There were associations with the sacred langur monkeys who were frequently seen near the macaques. The sacred langurs are slender animals, colored white or pale gray, with black hands, feet, and faces. The rarer purple-faced langur was less evident in association with the macaques, as this langur spent most of its time in the tops of the tallest trees.

Dr. Dittus had invented an ingenious system of weighing the monkeys. A scale with a large dial was attached to a rope that went straight up and over a tree limb and off at an angle to a smaller tree about fifty feet away. Hanging below the scale on three chains was a shallow pan to hold grain. Scale and grain pan could be lowered so the pan was resting on the ground. When the monkeys came to get the grain, a more adventurous individual was the first to investigate. He was allowed to eat some of the grain, and this was noticed by the rest of the monkeys. Slowly the rope was pulled to raise the pan above the ground. If only one monkey was in the pan, the weight was noted and the weight of the pan subtracted to give the weight of the monkey. This was recorded by Dr. Dittus, and over the years he had established growth records for many individuals.

The monkeys were accustomed to Dittus and soon became used to us as well. We filmed the monkeys resting in the trees, some of them sleeping, with the young playing nearby. Sometimes a young monkey would jump from a limb below, grab onto an adult's tail, and swing back and forth as if on a rope. The adults were very tolerant about this. The young were also interested in associate producer Peter Drowne's camera and tripod. While he was taking pictures, a youngster climbed up the tripod legs high enough to look in one end of a lens.

Dr. Dittus, after years of observing these macaques and their social structure and behavior, counts as most important his discovery that males, after maturing, frequently leave the troop into which they were born and become wanderers. When approaching another troop a wanderer stays on the periphery of the group, gradually moving closer and eventually getting acquainted with individuals. He is sometimes accepted as a member and becomes a breeder in the adopted troop. Other times a wanderer may be driven off by the larger males.

Females, on the other hand, never leave the troop into which they are born. Unrelated males joining a new troop bring new blood and unrelated genes and in this way minimize inbreeding.

One evening we were all sitting on an open veranda watching a sunset through some small trees. After the sun had set and it was nearly dark, Dr. Dittus pointed out a small tarsier-like primate running quickly down a sloping limb. It was silhouetted by the glow still left in the sky. "There is a slow loris," Dittus said. "We often see them here." I could hardly believe that fast-moving animal was a slow loris, for the ones I had seen in captivity were, as the name indicates, deliberate in their movements. I was able to see the animal well enough through my binoculars to be sure it was indeed a slow loris. On another evening we heard and saw with our flashlights other fast-moving "slow" lorises in some small trees. Here was another example of how the actions of an animal in nature may differ from those of the same animal in captivity.

We filmed in different sections of Pollonnaruwa, and Peter and I were able to take many still pictures of remnants of the fascinating carvings. We admired the highly carved semicircular "moonstones" at the bottom of steps leading up to the ruined palace and the round-topped guard stones with statues of humans in graceful poses at the entrances to shrines. Chronicled history was carved into a gigantic stone Ola Book twenty-seven feet long, four feet seven inches wide, and weighing about twenty-five tons. It came from a quarry sixty miles away. We also were able to get pictures of lizards, including two kinds of large monitor lizards, frogs, toads, and many different birds.

Our filming finished, Peter and I were ready to depart for Colombo. We took advantage of an extra day to drive slightly out of our way to visit Sigiriya, a fortress in the sky. Its sheer walls rising 600 feet above the plain, this remarkable ruin was built in the sixth century

A.D. We climbed to the top by a succession of steps and sloping ledges. A guardrail keeps visitors from sliding off and falling hundreds of feet to the rocks below. We had heard about the rock paintings under the overhang of the cliff, but I was unprepared for the quality of the artwork. These frescoes in the "pocket" are said to be the only ancient paintings in Sri Lanka that do not have a religious significance. Although the nearby graffiti mentioned "five hundred golden-colored ones," there were only twenty-two figures of women, mostly in pairs. The artist is unknown, and the paintings' significance is disputed.

Climbing further, Peter and I came out onto a broad terrace affording a marvelous view of the surrounding area. Looking down in one direction we saw an ancient city laid out with a long mall and ruins of numerous buildings. Steps led higher between two gigantic paws of a brick lion. Formerly the huge body and head had been connected with the forelegs and paws, but there was little left to indicate that. Instead, the steps petered out, and carved into the living rock were depressions for climbing, with an iron handrail to steady us. We worked our way up the dangerously steep ascent until finally we stood on the very top and wandered over the four acres of ruins. Little was left of the many structures once present, but we saw a huge ancient reservoir for water storage and, a stone's throw away, a ruined edifice that faced the rising sun.

Warren Garst and I had already filmed the story of Dr. Ray Carpenter's studies in Japan on the social and community organization of red-faced macaques, relatives of the toque macaques of Sri Lanka. Unlike their Ceylonese cousins, the red-faced macaques are heavily built, with longer muzzles and very short tails. We spent two weeks with Dr. Carpenter, dean of primatologists, filming these monkeys in their habitat on the small mountain called Takisakiyama on the island of Kyushu. Japanese primatologists maintained a feeding station for the hundreds of macaques on the mountain and had discovered much about their interrelationships.

The monkeys were divided by their own natural selection into three troops, named by the scientists A, B, and C. These troops knew each other, if not intimately then at least by the fact that members of another troop were recognized as being "them, not us." This engendered a certain survival technique of being alert, while feeding, to the approach of another troop. Troop A was by far the largest, but even so its members would rather retreat gracefully from the area at the end of their feeding period than have a showdown that might lead to an intertroop battle.

Sweet potatoes and grain were spread over the feeding ground. The status of individuals then became apparent, for when the alpha male approached a food source, others feeding there moved off. Female alphas dominated other females and lesser males. Each animal knew its place in the troops and responded accordingly when near others. Each individual dominated subordinate animals and was subordinate to more dominant members. A stare, protrusion of the face in the direction of a subordinate individual, or an open mouth was usually enough to assert dominance; if not, there was an attack and a chase, perhaps ending in a bite. Feeding continued until hunger was satisfied or the food was gone.

Grooming then began between individuals, usually being offered by a subordinate individual to a more dominant one. Grooming consisted of spreading the hair with the hands, looking for and removing particles of food or any bits of foreign material. This activity was initiated by the smacking of the lips to indicate a willingness to groom another monkey, a signal frequently repeated during the grooming. Monkeys rarely become infested with lice or other insects, but if these are found in the hair or attached to the skin, they are removed and eaten by the one grooming. A monkey who has been grooming sometimes is groomed in turn by the other. Grooming is an important part of the social structure of these and other monkeys.

Around the edges of the feeding area, young animals were playing, chasing each other, carrying on mock fights, rolling over and over on the ground, or playing a version of King of the Mountain.

When it was about time for the next troop to arrive, the adult males became alert and watching. As soon as the incoming troop was spotted by a male, he quickly climbed a tall, slender tree, and taking hold of the small trunk near the very top, shook it violently until the tree swung back and forth with the movements of his body. He gave a warning call as well. Then the whole troop moved off in the opposite direction from the approaching troop. Those on the ground stood up to get a better view of the approaching monkeys, and vocal communication continued to alert the troop and keep it together as they moved off into the forest.

When we had finished filming at Takisakiyama, we went by car and boat to a small island just off shore called Koshima. There lived another group of Japanese red-faced macaques. We landed on the sandy shore of a cove, and as we did the monkeys came to meet us. We had brought sweet potatoes cut into small pieces, and these were spread out on the sand not far from the water. A female macaque was pointed out to us as having been the first to discover that she could get rid of the sand sticking to the potatoes by taking pieces to the water and washing off the sand. The others, seeing her so easily and quickly dispose of the sand, copied her actions until almost all the troop waded out into the shallow water to wash their potatoes clean. "Learning by observation" is a term often applied to this ability in monkeys, but the old adage "monkey see, monkey do" says it just as well.

25

Lapland: The Great Reindeer Drive

Everyone connected with "Wild Kingdom" contributes show ideas. Before we became "Mutual of Omaha's Wild Kingdom," Don Meier and I made up a list and a rule: The only way a show idea could come off the list was to be filmed.

One of the ideas I had suggested was a story about the Lapps and their reindeer. I had long been intrigued by the semidomestication of reindeer by the colorful Lapps and by the long journeys north in the summer and south in the winter, as the Lapps move their reindeer to the best feeding grounds. It was not until fifteen years later that Titus Vibe-Mueller, a Norwegian film producer, showed Don a film of the Lapps and their reindeer herds on migration to the northern part of Norway. From this meeting a plan was worked out and a date was set for us to film "The World of the Lapps."

The Lapps probably originated in Central Asia and gradually moved northwestward to their present Lapland. Lapland is not politically a country, but covers part of Norway, Sweden, Finland, and Russia. The Lapps are small people, averaging five feet in stature, but are strong and muscular. They live above the Arctic Circle, where the birches, pines, and firs are stunted because of the cold weather and short growing season. Of the thirty thousand Lapps thought to be alive today, twenty thousand dwell in Norway.

On April 10, 1976, I left JFK airport in New York on SAS's 5:05 P.M. flight for Copenhagen, bags tagged to Oslo. I had hoped to get three seats across so I could lie down and sleep, but the plane was full and I had to doze sitting up all night. The next morning Copenhagen was fogged in. We circled overhead awhile but had to divert to Stockholm. From there another flight took me to Oslo, but because I was late no one met me. A short taxi ride set me down at the Globetrotter Hotel, where Warren Garst and Peter Drowne, the photographer from Los Altos Hills, California, were waiting. They introduced me to Titus Vibe-Mueller, who was in misery with water on his knee and could scarcely walk. Despite this, we all went out to dinner and I was brought up to date on the activities so far and the plans ahead. Titus could not accompany us on the trip as he had hoped, but had arranged with Hans Hvide-Bang, the photographer of the film he had shown Don in Chicago, to meet us at Alta, a town in the far north.

Next morning our plane left Oslo at seven, and after a stop at Tromsø, flew on to Alta, where Hans Hvide-Bang was waiting. He had engaged as cook Walter Holt, a large man with a full black beard who had traveled the world as cook on Norwegian ships. We went into town to shop for skis and boots and other necessities. With our equipment and luggage distributed in Hans's VW station wagon, Walter's Ford station wagon, and a rented trailer, we started our drive to Kautokeino, headquarters village of the group of Lapps we would be working with. Three flat tires delayed us, but we finished the seventy-eight-mile drive in time to check into the Tourist Hotel for dinner.

Kautokeino lies in a valley below the hotel. A picturesque church on the far side and the many bright-colored houses against the white snow make it a picture-postcard view. There was an air of excitement in the village that day, as the Easter season was nearly upon us. The Lapps were in their most colorful costumes. Hans introduced us to some of the families we would be traveling with. At one of the neat little cottages we were given tea and cookies and saw that the Lapps live much as the Norwegians do, with nice furniture, curtains, rugs, and pictures on the walls. They drive cars and are involved in other businesses besides reindeer.

We spent three days getting snowmobiles, sleds, food, and other essentials organized for the long trip north. The herd had been moved out a few days before and was feeding at a Lapp encampment where reindeer moss grew. Some animals were still in town for the annual Easter reindeer race, which was attended by a large crowd and which we filmed. Our guide, Nils Logje, a handsome twenty-two-year-old Lapp, and I watched near the starting line as several teams, each composed of a reindeer, a sled, and a driver, were given the starting signal. A great shout arose from the crowd as the drivers urged their reindeer forward. Some drivers ran behind to allow greater takeoff speed for the reindeer and then jumped on the sled, trying to get as much speed from the reindeer as possible. They went around the circular track, which was marked by stakes driven into the snow. Some animals were hard to guide, ran off the course, and were disqualified. One turned right instead of left, broke its way through a line of spectators, and scattered them, much to the amusement of onlookers nearby.

The men raced first, then the women, and finally the children. Food and soft-drink stands did a thriving business. Nils became bored just standing there to be photographed for our movie, and during a lull between races he slipped off to have some fun with his friends. At the end of the competition, trumpets summoned attention, prizes were handed out to the winners, and many photographs were taken. It was a fun day.

Easter Sunday I walked to town with my still camera to shoot pictures of the village and its people. I found myself at the church at eleven o'clock and went in for the service. Most of the worshippers were in their most colorful Lapp clothes; there was only a small sprinkling of regular suits and dresses. The service was in the Lapp language. Some of the music was familiar, but I had difficulty following the proceedings, so I slipped out and waited for the congregation to leave. As they departed I took a few photographs, as picture taking was not allowed inside.

That afternoon I placed an overseas call to Saint Louis and had a nice Easter visit with Carol. It would be our last contact until after my 150-mile journey with the Lapps and their reindeer, for there were no phones or radios in the mountainous country through which we would travel.

Soon afterward I left the hotel to join the others at the starting point, the home of one of the Lapps on the edge of town. All our gear, camera equipment, food, propane gas, stove, reserve supplies of gasoline, and other necessities were packed on sleds, covered with reindeer skins and canvas, and securely tied down. Each snowmobile pulled a sled. We had four units. Nils led; I followed him; then came Warren and finally Peter. Hans could rent no additional snowmobiles so he rode astride the pile of equipment on a sled, and his fourteen-year-old son, Hans Peder, rode on another. The trail led over rolling hill country, frozen lakes, and scattered birch woods. The going was rough, and I kept slowing for bumps. After a time I realized I was getting farther and farther behind Nils, so I put on some speed to catch up and paid less attention to the rough places.

Nils stopped about every forty-five minutes so we could stretch our legs. At one of these stops Peter did not show up, so Nils went back to find him and lead him on. Nils returned alone, and after a conference we decided to continue, hoping Peter would find our trail and catch up. As the light faded into dusk the last part of this day's journey was lit with our own headlights. We had left Kautokeino at 7:30 P.M. and at 11:00 we arrived at a place called Denomoski, a wooden hut with four bunks and a stove. We would spend the rest of the night there.

Nils and Hans, however, went on to the Lapp camp, about one-half hour away. They returned with Peter, who, having stopped to retie the ropes on his trailer, had somehow lost our tracks in the snow. He must have turned a little to the right, for soon he found other tracks which he followed to the Lapp camp. What a relief it was to see him! All's well that ends well. We were in our sleeping bags about 12:15 A.M.

After breakfast the next morning we left at eleven and stopped at the Lapp camp. I saw their tents for the first time and was reminded of Indian teepees or wigwams. Long slender poles were spread at an angle and protruded above the blankets surrounding them, forming a cone-shaped dwelling. Years ago the Lapps used reindeer skins sewn together, but blankets are lighter and easier to pack. The tent had an entrance with a flap covering it to keep out the cold. Smoke emerged from an opening just below where the poles converged.

We were invited into Nils's home to meet the Logje family. Nils's father, mother, and sister sat around an open fire on reindeer skins and blankets and had coffee and dried reindeer meat which was toasted in the fire on the end of a stick. We took some pictures inside with these laughing and vivacious people.

Soon we were on our way, for we had a long distance to cover to reach the next government hut. After a time it began to snow and the weather deteriorated. All at once we were in a whiteout. Following Nils at thirty to forty feet, I suddenly could not see him. He and his snowmobile and sled had disappeared. I followed his tracks and he reappeared. But not for long. We were traveling more slowly. Then because we could see nothing in front of us or around us, we had to stop.

We sat in the driving snow for an hour before the whiteout lifted and we could proceed. The wind was picking up, and we were grateful when, just ahead, we saw our cabin. The snowmobiles and sleds we parked in the lee of the cabin and moved what we needed inside. Clearly a blizzard was building, and we would be here for a day or two until the storm blew itself out.

The cabin was twelve feet square, with a small vestibule on the side where it would get the full blast of the wind. There were two double-deck bunks, a stove, a table, and three chairs. There was one small window which we soon covered up to keep out the cold. Nils and Hans took two snowmobiles and sleds to get firewood, as there was none in the cabin. It was a difficult trip. One of the snowmobiles conked out and they had to leave it. They returned with one sled piled high with wood. We were to need it, for the temperature dropped, and we were only really warm when we were in our sleeping bags. Warren, Peter, and I had a bunk apiece. Hans and Hans Peder shared the fourth. Walter slept on the table. Nils slept on some reindeer skins.

The howl of the wind increased and snow came in around the door. We placed reindeer skins and blankets around the doors of the vestibule and the cabin. We made the hut as tight as we could, but the snow came in anyhow. Water froze in the cabin. Walter cooked a good meal of lamb and cabbage. When I crawled into my sleeping bag my feet were like ice. I took off my socks one at a time and hand-warmed each foot, then put the sock back on, an operation which in a mummy-type sleeping bag is a very neat trick. I slept in my underwear—two sets of long johns. The wind howled all night.

The Lapps were scheduled to break camp the next day and move on north. The blizzard increased in the night, however, and blew all the next day. Warren Garst had with him a pocket wind gauge. During the afternoon when the wind was screaming, he bundled up, went outside, and exposed the gauge to the wind. The indicator went to the top with such force it broke the gauge. The top mark on the gauge was sixty miles per hour. The gale almost blew us over when we had to go outside.

At the end of the third day the storm abated. The snow stopped; the wind lessened; later the sun came out and the air began to warm up a bit.

Nils went back to the Lapp camp to find out about the reindeer, for he was worried that some might have wandered off or become separated. When he returned he said one of the tents had been blown down, but no one was hurt. Some reindeer had indeed become lost in the whiteouts and these had apparently dispersed. The plan was to push on with the remaining animals, leaving some of the Lapps to scout for those that were lost. The Lapps said this was the worst storm in ten years.

We spent a whole day in their camp filming their activities, including lassoing the calves and notching their ears with each owner's brand. Each owner of reindeer has his own distinctive ear notch as a brand, so he can always recognize his own animals. All the brands are registered with the government, and no one is allowed to use another person's brand. Each person notching calves saves one piece of cut-out skin from each calf and strings them together to form a record of how many calves have been added to his herd. Dogs help in the roundup by driving the reindeer past the lassoers, who await the chance for a good throw.

Later in the day we filmed the breaking up of camp, which included striking the tents, loading them on sleds, harnessing the draft reindeer to the sleds, forming a column, and moving out toward a new camp. At the head of the column a woman led a draft deer pulling a sled. All fell into line behind, and the dogs helped to control the loose herd and keep the deer in their places during the march. As reindeer strayed away from the main group, the outriders on skis and the dogs would turn them until they were back with the herd. It was most interesting to me to see how gentle and careful the Lapps were with the deer. When putting an animal into the shafts of a sled, one man raised the shafts high, while another moved the deer backward or sideways with a gentle guiding motion of hand or halter. No force or rough handling was used.

We leapfrogged ahead to position cameras to film the column, then hurried ahead to film the march at another location. At one point two of our snowmobiles were commandeered by two Lapps who dashed off to keep a band of about seventy-five reindeer belonging to another group from getting mixed in with their own larger herd.

They were successful in this and soon returned our snowmobiles. We moved forward again and continued to film the advancing column. Some forty sleds were in the march, along with 3,000 reindeer and a number of people walking. Other people were on skis, and the dogs kept the herd together and moving forward. Each camp was on a hill marked with a cairn of rocks, some on top of the pile painted yellow. These were places where reindeer moss grew, and the animals could have the food they required. The camps were about six miles apart.

At another campsite on a hilltop there was no wooden hut, so a family of Lapps pitched a tent for us and provided wood for our fire. Above the fire in the center of the tent a length of chain held an iron pot. A rope was attached to the top of the chain to raise or lower the pot. Reindeer skins were spread around the inside edge of the tent. There we could place our sleeping bags and eat and rest while leaning against our duffel bags. The outside of the tent was weighted down with heavy stones as a high wind was blowing and snow was falling.

The Lapps came by to see how we were doing, and one woman, who disapproved of the skimpy door flap, came back with a much better one. The top hole in our tent was bigger than the others, but even so we were bothered by the smoke and lay down when we could to breathe the fresher air near the ground. Walter, the cook, set up his propane stove to thaw and warm up a precooked stew that had been brought along in a large plastic box. He chopped off pieces of frozen stew and heated them in a large cooking pan. The stew, with thin flat-bread waffles, was our supper, along with rye bread and jelly. Soon afterward we were in our sleeping bags.

At five in the morning, as I crawled out of my mummy bag, still clad in two sets of long johns and wool socks, I found my other clothes covered with snow that had blown in during the night. After shaking off the snow, I put on my wool shirt, thermal pants, more wool socks, snowmobile boots, red down jacket, muffler, wool balaclava, snow-mobile suit, and mittens. We filmed some departure scenes for which I had to take off the suit and balaclava and put on a red headband. A black suit doesn't do much for a scene against snow. After filming, when we were ready to travel, I quickly slipped back into the suit and balaclava, and added my helmet.

For those from the warmer parts of the world who know nothing about snowmobiles, they are best described as similar to motorcycles. To ride them you sit astride a seat with your feet on small platforms, just as on a motorcycle. In front is a gasoline engine, which is started by a quick pull on a small rope. The driver is protected from the wind by a curved plastic shield—a windscreen. The snowmobile is guided by handlebars and attached to two short skis. You push on a throttle with your right thumb to go faster. As the motor increases in speed a clutch is automatically activated, and that starts a drive shaft which rotates cogwheels that move the wide, cross-cleated rubber tread upon which most of the weight rests. There is a brake on the left handhold.

When you travel over snow and ice in cold weather, the wind-chill factor increases with the speed. The rider needs a warm wind-breaker and snow-repellent clothing, boots, headgear, and mittens. In subzero weather I wore a wool balaclava tucked in around my neck, with the thermally treated hood of my snowmobile suit over that. A padded, hard-plastic helmet with chinstrap and plastic visor went on over that.

The snow increased as we approached the column of sleds and reindeer. Then it turned into a whiteout and we had to stop. After a long time the air cleared enough so that we saw the herd start up the broad side of a mountain. We followed in the high wind and blowing snow. Before long we were all stuck in deep snow on a considerable slope. Nils was the only snowmobiler with enough experience to get us out of that situation. He was in the lead testing the slope and the snow. He would zigzag in a series of cutbacks, always turning the snowmobile downhill a little when it began to bog down, then picking up speed to keep going. He was adept at standing and leaning out on the uphill side of the machine in order to maneuver it into a turn and get the moving belt to bite into the slope.

He too got stuck. But he was a strong young man, and he lifted and hauled his machine into a downhill position. Then he got the heavily loaded sled as close to the back of the machine as possible, and standing alongside and pushing on the handlebars, he revved the motor until the machine moved forward free of the sled; then the linkage took up, and as the snowmobile gathered momentum he jumped on and was again in motion.

I don't know how many times Nils rode his snowmobile up to a slight saddle, disconnected the sled, and rode back to get one of us going. Sometimes he would ask one of us to ride his machine and follow him as he drove our unit to a saddle. Then with his machine he would return for another and then another.

There were four riders altogether—Nils, Warren, Peter, and me. Hans was riding whenever he could and sometimes walking. I was exhausted, cramped, hands and arms aching from manhandling a stubborn snowmobile and trying to pull a heavily loaded sled up a mountainside in heavy, deep snow. I wondered if I would ever make it. Hans noticed and kindly offered to drive my unit. Although I was slipping and stumbling in the snow, walking was nevertheless a relief from hours on the snowmobile, and I kept plodding upward. At this time Peter and Warren were stuck, and so was Hans. Nils finally got his unit to the top, separated his sled, came back to where Hans was stuck, and with his lasso tied his machine in front of Hans's. Following Nils's own tracks and with both engines pulling hard, they moved forward and upward to the top. Then both snowmobiles returned and with a second lasso, both machines in tandem in front of Peter's machine, and with all three engines running, they got Peter to the top. By this time I had walked in the deep snow between the whiteouts to the top.

There were times when I thought we wouldn't make it without additional help. But I underestimated Nils's ingenuity. With the aid of ropes and with one or two machines in tandem, he finally got us all to the summit. Ahead of us was level or downhill travel. After a brief rest we were ready to push on. Nils's dog, who had been with us since leaving camp, looked tired too. Nils invited him to jump on top of his trailer, where he rode the rest of the distance.

The importance of traveling close together was proved on the downhill trail. Three units jack-knifed and slid off the trail to the left. It took all of us to get them back on the trail and moving forward.

Eventually, tired, cold, and wondering what would happen next, we arrived at the Kildal hydroelectric station, our first destination on this long and exhausting day. The manager invited us into his snug house for coffee. His wife had been with one of the families on the migration and had arrived with the first group of Lapps ahead of us. Coffee, sandwiches, and doughnuts tasted mighty good, and it was a relief to get out of our heavy outer clothes and our boots and be warm. I was also happy to learn that this was the end of our snow-mobiling. The Lapps would take over our machines and use them to round up the stray reindeer. Hans called the Hotel Sorkjøsen, on a fjord near Kägen Island, to make sure our room reservations were in order and then called a taxi with a trailer to fetch us and our gear. The manager assured Hans that a car was ready to leave and could take the three Americans to the hotel, about nineteen miles away. Warren, Peter, and I gratefully accepted and, with one suitcase each, arrived at this nice little hotel at nine in the evening. Dinner was being served until ten. A quick wash and clothes change, and we were in the dining room eating everything in sight.

One of the rewarding experiences of on-location filming for "Wild Kingdom" is the comfort of arriving at a good hotel afterward. To be able to push a button and see a light, turn on a faucet and get water at the temperature you want for a shower or a tub bath, wash, dry yourself with a clean, warm towel, and have a large mirror for shaving is bliss. There was a flush toilet, a telephone to put you in contact anywhere in the world, and radio and TV, carpets, and good beds. An added bonus was a fine view of the fjord.

Next morning at breakfast we learned that the main herd of rein-deer and the Lapps with them had arrived at Kildal at 11:30 P.M., perhaps more tired than we. They made camp by the lake and had still another day's walk to reach the place on the shore of the North Sea where the reindeer swim to the islands.

The women and children left the migration at Kildal, as did some of the men. Only a few made the walk with the reindeer to the coast. Fewer still stayed with the herd on the islands during the summer.

We spent a day sorting and checking equipment, recharging the camera batteries, reloading magazines with film, and making plans to shoot the swim to the islands. We called the office in Chicago and our homes to let them know where we were.

After another good night's sleep we were up early and ready with our cameras to film the swim across to the islands. A truck took us to a pier where a boat met us and transported us to the site where we were to meet the Lapps and their reindeer. But they had not arrived. We waited, but they still did not come. Finally we were put ashore near some houses, and Hans called for a taxi to take us back to our hotel. He then made arrangements for a local man to snowmobile into the mountains to try to locate the Lapps. He carried one of our walkie-talkies and was to contact us each half hour. On the first contact we received a message that the Lapps and their reindeer had met another blinding snowstorm and the men had lost contact with the herd. They were now trying to find the animals.

The second communiqué brought the good news that the herd had been located and was being moved to a grazing area on the mountain. The Lapps had given up hope of moving the animals to the islands that day. The march probably would be delayed two days.

Late in the afternoon one of the Lapps came to the hotel and told Hans that they had located only half the herd. They were scouting for the others but the snow was very deep. Other scattered bits of information trickled in to us. The Lapps might leave the animals where they were if the grazing conditions were good and the reindeer could get the moss in the deep snow. They might remain where they were for another twelve to fourteen days, as some deer had already calved. In that case we would have to remain, for our film would not be complete without the swim to the islands. We must hang loose.

During this waiting period, I had some free time. In my hotel room I spent the time writing some of the early chapters of this book.

The Lapps were camped on a mountain top above the hotel. Although only one snowmobile was available, we managed to get to the camp with a camera and tripod. The ascent was too steep for the machine to pull a loaded sled, so some of us walked. Several days were spent filming the Lapps' camp activities—the erection of tents, called by them *kata*, and the method of taking them down and stowing them on the sleds. We filmed Lapp movements on skis and their use of a lasso to rope a draft reindeer, the harnessing of the animal to a sled, and the manner of driving and guiding the deer with one rein.

The Lapp lasso, *suohpan*, is a smaller and lighter rope than that used by American cowboys. It is also thrown differently. A Lapp holds the coil in his right hand; the rope end and the bone slip *(kiela)* are held in his left hand with a small noose stretching back to the right hand. The coiled loops, the small noose, and the bone slip are thrown together and the end held in the left hand.

One afternoon we practiced, after being shown how to coil and how to hold the noose and the coiled loops. After a time we could throw the lasso, but not nearly as well as did the Lapps.

We made many trips to the mountain top to film the Lapps and to find out about their plans, particularly those related to the swim to the islands. It was frustrating to spend day after day without knowing the full situation. It must have been frustrating for the Lapps too, but they were more stoic than we; the reindeer were their chief concern. While scouting for the part of the herd that was dispersed, they worried because this mountain did not have as much reindeer moss as most others and because calving time was near. They wanted to get the deer to Kägen Island, a swim of half a mile, where some could stay for the summer. Other animals could swim from Kägen to Arnøy Island, a distance of one mile, but only after the deer had rested and fed for a couple of days on Kägen. They hoped to get to the islands before the calves were born, as the babies could not make the long swim. The delay in the blizzards had given the Lapps reason for some anxiety. The reindeer were their livelihood, and to lose calves was a serious loss.

At last we got word that the swim would begin. We were waiting on Kägen Island with two boats, watching the mainland with binoculars. A holding pen for the animals was set up at the water's edge with the side toward the sea open. Usually the reindeer had to be forced into the water. An opening in the back gave access to the enclosure.

We jumped into the boats, cameramen in one and Nils and I in the other, and started across. The animals entered the enclosure and milled around. When we were within about 100 feet of the shore and before our cameras were ready, a reindeer, without any encouragement, walked into the water and started to swim. He was followed by another, then another, and before our cameras were ready to film, the whole group of about 150 animals were in the sea swimming to Kägen Island. We moved our boats in the same direction and came in behind and to one side, but the herd was moving rapidly and we did not get a foot of film. What a disappointment!

Two days later we were assured by the Lapps that they had rounded up the animals on Kägen and were holding them in a pen ready for the mile-long swim to Arnøy Island. To make sure we would be there early the next day, we stayed overnight on Kägen in makeshift quarters and were ready to leave the dock by five in the morning.

We had to go by boat and it was a long ride, but we arrived and started to film in the large holding yard by eight. Yohan, Nils's father, explained to me that he would lead a draft animal to the boat at the beach and that a kinsman, Anslak, would be on hand to give the deer a push from behind if necessary. Nils would start the outboard motor and I would be at the back of the boat ringing a reindeer bell by hand (the bell would make no sound if left around the neck of the lead reindeer in the water); the other animals, we hoped, would follow by sight and sound.

When everyone knew the drill and cameras were in place to film from three positions, Yohan threw his lasso and caught the lead draft reindeer. He and the deer and Anslak and I moved toward the boat. When we were nearly there another Lapp removed the burlap strip from the side of the pen nearest the sea to make a large opening. Yohan got into the boat and pulled on the lead line. Anslak gave the deer a gentle nudge forward, then jumped into the boat. I began ringing the bell as the deer entered the water. Then I quickly jumped into the boat myself, while Nils pushed off and started the motor. The lead reindeer was soon swimming at the end of his rope behind us. The other Lapps urged the animals forward from the rear of the holding yard. I had wondered if they would really follow and was greatly pleased when they did. Looking back, ringing the bell, I saw group after group walk over the slippery rocks of the shore and move into the bone-chilling water of the Arctic Ocean.

They swam with their heads and antlers held out of the water. There was a small bow wave where the short neck came out of the shoulders, and the hollow hair gave them buoyancy. Another boat pushed off from shore to bring up the rear and to help any stragglers. Our photography boat moved to one side of the swimming herd but not close enough to bother them. While Warren filmed the swim, Peter's boat sped across to the landing site, where he filmed the arrival.

Everything went well, and we landed on the rocks of Arnøy. There Yohan released his lead animal, which moved on up the bank, followed by those behind. Each animal shook itself, releasing a large spray of water, then moved off at an angle up the mountain. They were heading for a valley not visible to us.

Anslak and I went back in the boat to look after the stragglers. Some were slow, but all made it to shore. One small animal needed help. I took hold of its antlers and held its head up as we towed it to shore. Yohan was waiting when we landed, and together we got the deer ashore, rubbed the water from its hair, and massaged its legs and body to get the circulation going. After a time it was rested enough to want to stand, and then it slowly moved up the slope to join the others. Its walk became stronger as it moved off behind them. Watching through binoculars, I saw it trotting to catch up. I knew it was all right.

26

Crocodiles of
Papua New Guinea

When I was a young curator of reptiles at the Saint Louis Zoo in the early 1930s, most crocodilians were abundant. Florida was doing a thriving business in the sale of baby alligators as pets, the Nile crocodile in Africa killed more people than any other vertebrate, and caimans from South America were beginning to be sold in Europe and America in competition with the alligator. Only a few species were considered rare. None seemed to be endangered.

This was soon to be changed. Improved methods of tanning the skins and processing them for the leather trade increased the demand, and laws were required to save the rapidly dwindling numbers of alligators in Florida. Prices continued to rise, and alligator poaching became a way of life for many in the Everglades. The American crocodile, never abundant in extreme southern Florida, was hunted almost to extinction.

Enormous numbers of skins found their way to dealers and processors, and it became the vogue to have a pair of alligator shoes, a crocodile handbag, or a caiman traveling case. The attrition of the crocodilians was worldwide.

When we heard of the efforts being made in Papua New Guinea

to establish crocodile farms and permit only a controlled sale of skins through a government agency, we were interested. Don Meier Productions contacted Jerome Montegue, a volunteer of the United Nations Development Program, at his crocodile farm on Lake Murray, and Montegue invited us to film an episode there. Logistics were worked out, and before long Carol and I were on our way.

Associate producer Peter Drowne had preceded us and was on the airstrip to film our arrival at the Lake Murray crocodile station. Jerome Montegue was there also, a handsome young Kentuckian. We were shown to a building on stilts about twelve feet above the ground, where we were to stay. The New Guinea sound technician, Peter Lakome, with whom we had worked on a previous trip to Papua, and Peter Drowne were quartered in a prefab building not far away with room to store and service the photographic and sound equipment. Water for our cabin was hand-pumped to a storage tank on the roof directly from rainwater storage tanks on the ground. The weather was very hot.

After a quick lunch Jerome, Peter Drowne, and I visited the enclosed yards where the crocodiles were kept. Jerome outlined the program to conserve them. The government had passed a law protecting animals larger than twenty inches across the belly. (The measurement is taken from just behind the front leg at the edge of the ventral (belly) plates across to the other edge behind the opposite leg.) Up to this size they could legally be killed and skinned. It was forbidden to kill larger ones because then they were mature and had to be left in the wild to breed. When crocodiles in captivity reach this size they are liberated to become breeders. Baby crocs become more valuable as they grow. This is an incentive to keep them and let them grow, for then their larger skins can be sold for more money. To ensure protection, all skins exported from the country had to pass through the crocodile skin department of the Papua New Guinea government. Near Port Moresby two other large crocodile-breeding farms had been established, and were successfully raising young animals.

Jerome had three crocodiles that had grown in the Lake Murray crocodile farm to a size that mandated release. We filmed the release next day at a suitable place on the river. We traveled there by dugout canoe, and Jerome and I freed the crocodiles while Peter Drowne filmed from a river truck. These crocodiles were over five feet long, and their struggles to be released made quite a show.

The river truck, a flat-bottomed scow with two outboard motors, was to transport us up the Strickland River to contact native crocodile hunters. This was an important part of the crocodile-conservation plan. As leader of the project in this district, Jerome made trips as frequently as he could to talk with the natives about the regulations and about buying crocodiles. He had to have two interpreters, one who spoke English and Motu and one who spoke Motu and the language of the village visited. Our trip was to take five days. That night we packed our gear along with several pounds of black stick tobacco to pay the natives for their help. We took kina, the money of the country, to pay for the crocodiles. In addition to kina and stick tobacco, newspapers were a part of the price, as they were used for cigarette paper to roll the tobacco in. A national law recognized this and prohibited newspapers from using lead-containing ink.

Next morning, with all our gear loaded aboard the river truck and two large dugout canoes, also with outboard motors attached, we pulled away from the dock, as Peter Drowne recorded it on film. The river ran through a jungle that came right down to the edge of the water. The thick, dense growth was difficult to get through. We saw many birds—flocks of cockatoos, egrets, herons, ducks, and magpie geese. Then there were beautiful bell magpies, white with black head, tail, and wing-tips, and with a bell-like call.

Before long one of the river truck's motors conked out, so Jerome asked the two much faster canoes to slow down and stay behind us in case we had more trouble. When we reached a point where the Herbert River flows into the Strickland River, we turned left to go upstream. As we did so, our remaining motor stopped in the thick mud of a shallow tongue. Peter Drowne put an oar out to hold us and let the current turn us away from the mud. But, when the engine started, we almost lost Peter. Just in time, he let go of the oar, and we had to circle around to pick it up.

Our speed was reduced by the current we were bucking, and we passed through many rain squalls on this five-day expedition. At one bend we came upon nine black-and-white Australian pelicans; they glided across the river close to the water, flapping only when they needed more speed, a flight pattern typical of pelicans.

We stopped at four temporary villages that had been built along this part of the Strickland River, each provided with shelters resting on stilts and having walls of split bamboo. At each stop Jerome con-

sulted with the inhabitants about the regulations that had been devised to conserve the crocodiles. This was Jerome's fifth trip in two
years, and everywhere we went he was greeted by friends. For the
villages along the way, his visit was almost their only contact with the
outside world, and his friends were excited and happy to see him. At
Tanukomal, the farthest village, we spent an entire day filming in and
near the river. We photographed Jerome buying, measuring, and paying for crocodiles. We also filmed the activities of the natives—their
small vegetable gardens, their canoes traveling on the river, their
spears and knives and weaving. But the emphasis was on crocodiles
and what their conservation would mean in these remote places,
where crocodiles were almost the only cash crop.

We filmed clear through the day, and when the light dimmed, we
were tired and hot. We men went down to bathe in the river, while
Carol bathed from a five-gallon can of river water in our house on
stilts. She poured the water over her and it ran right through the split-
bamboo floor and onto the ground below.

Next morning we were up early to begin our return trip. All
forty-five members of the village of Tanukomal turned out to help
carry our things to the boat, pack them in place, and wave goodbye.
Over the big pile of equipment we spread a tarpaulin and tied it down
to keep the gear dry, for it rained off and on every day. Carol kept a
journal, which gives a pretty good idea of river travel on this expedition. She wrote of this return trip: "It was really very cozy traveling
down the winding river, with jungle walling us in on both sides, and
no other human being for miles and miles. Even our support canoes
were far ahead. They were supposed to keep us in sight for security in
case our motors should fail. There were two spare motors and extra
gas in the canoes, and we might need them. But the long, sleek dugout
canoes can travel much faster and the natives love speed, although
going at full throttle uses a lot more of our precious gas.

"When we first left the village our boat was alive with flies. Even
the wind didn't dislodge them. I had a can of Black Flag with me, and
so that the wind would carry the spray back to the flies instead of into
my face, I crawled over the gear to the front of the boat. Then I began
my attack. To get out of my line of fire, the three men jumped to the
side of the 'truck,' standing on the edge with their heads above the
pile of equipment. When they grabbed the tarp to hold on, the pockets of rainwater were released and they all got their stomachs

drenched. They looked so surprised, and it was so funny to see them leap to safety only to have this happen, that I laughed until I was crying, lying on top of the gear with tears running down my face. No one was steering and we were heading for the bank. Luckily it is a very wide river, everything was sorted out in time, and the flies were all gone.

"We had been traveling for ten hours. It was after six and getting dark. The shelter Jerome hoped we could sleep in was caved in and the platform not safe to walk on. Finally we came to the same shelter we'd slept in on the way up last Sunday night, the one we call Mosquito Camp because the insects were so thick you could hardly see through them. But it was at least a fairly solid shelter off the ground. It was getting very dark and we hurried to get our gear up on the platform while we could still see a little.

"Jerome started the lantern, but the primus stove didn't want to start. We had our three chairs from the boat under the platform, which was seven feet from the ground, so at least we could sit comfortably under the floor, with a wood fire. The smoke helped to keep the mosquitoes down. The tea kettle was boiling.

"For supper we had canned ham, sliced and warmed in canned margarine with sliced onions and mixed vegetables, and hot vegetable soup, all cooked one after another in the one pot that we could find tonight. The rest must be in the box that got put in the canoe. It was cozy there, all together with our little fire and warm food, safe and dry. I had the awesome feeling that within about 100 square miles we were the only people like us—just a few scattered villages. There seemed to be no living things except birds and insects on land, crocodiles and lizards near the river. We had seen only one snake and it was swimming.

"I stretched out the plastic dropcloths again, the men hung the mosquito nets, and I pumped up the air mattresses.

"It was raining hard but the mosquitoes drove us to our beds as soon as we had eaten. I had been in my muddy clothes since yesterday afternoon, so Marlin held up my sheet and I got into my pajamas (I have the 'riverview balcony' spot). Slipping between our sheets on our air mattresses seemed like heaven. Peter was playing one of his musical tapes very softly. The lantern made our little shelter seem like a safe island in this jungle.

"I had made a serious error of judgment. Because the jungle was so dark and it was raining so hard, I had taken a chance of getting through the night without going to the local powder room—the dark and drippy jungle.

"I awoke at about 5:00 A.M. just miserable with the need to go. The 'stairway' to the platform was one slippery log with no notches cut into it leaning diagonally against our platform. I had had quite a time getting up it in the first place, as my shoes were so muddy that Marlin had to help pull me up. There are large areas of our 'floor' that are not safe, with spaces big enough to fall through. I suffered, looking at that total blackness. It was a seven-foot drop to the ground. I had no idea where my shoes were in the piled-up gear outside my net. It doesn't begin to get light until six o'clock, but I was really in agony. I woke Marlin and whispered that I was going out and to hold my flashlight while I dropped to the wet ground. It was safer than slipping off the log. I landed okay and didn't venture far into the dark. I was nearly eaten alive in the few minutes I was out of my net. I couldn't get back up. Marlin held my hand and pulled, but my wet feet just slid off the wet log. I got one of our chairs from under the platform and put it next to the log. We had by this time awakened Peter, so he held the flashlight while Marlin hoisted me back to safety.

"There was no more sleep now as birds were waking up and dawn was coming. About 6:30 Peter started a fire and put the kettle on. After a quick breakfast of canned pears and coffee we packed up, buried our empty cans, and were off down the river once more."

27

Exploring the World of Sharks

Dr. Don Nelson of the State University of California at Long Beach had been studying the behavior of sharks. He had discovered that some sharks assume an agonistic attitude when threatened. They arch their backs, drop their pectoral fins below the bottom line of their bodies, elevate their heads, and swim in an unnatural and erratic way. To study this further, Dr. Nelson built an underwater submersible—a craft with the maneuverability of a submarine and powered by electric motors. Equipped with air tanks for breathing, an investigator could slide into this protective cage underwater and perhaps find out how much it takes to provoke a shark actually to attack.

Don Meier had been in touch with Dr. Nelson and in one of their conversations Don learned about the submersible and that it was to be tested at Eniwetok Atoll. American navy personnel were stationed there. Food was available in the mess hall, and we could sleep in barracks. Dr. Nelson's project was under the blanket of the University of Hawaii department of marine biology, and they would help us with arrangements for transporting equipment. When all the details had been worked out and Don Meier had suggested a story outline, Ralph Nelson and Rod Allin with all their photographic gear and scuba equipment went ahead to see the local situation and work out with Don Nelson and his graduate students the details of the story outline so that a filming schedule could be arranged.

After about a week I followed their route across the Pacific Ocean to join them. In Honolulu I checked in at Hickam Air Force Base with the Military Air Command (MAC) and was given my authorization to travel on a plane leaving for Eniwetok.

As I took my seat I was handed a small envelope containing two cones. A sergeant explained that this was wax to form earplugs to keep out the noise. Looking around, I could see that the belly of our huge plane was designed to carry heavy military equipment, and the bare metal of the shell was visible between the strong supporting girders. The forward part of the plane's floor was loaded with cargo for delivery to Eniwetok to support the military operations going on there to decontaminate the area where atomic bombs had been tested years earlier. The plan was to seal off those areas still emitting dangerous amounts of radiation, so that the original inhabitants could return to other parts of the atoll and live there. I had been assured that the areas where our filming would take us were perfectly safe, but as I remembered what this atoll had been subjected to, I had some second thoughts about going there. Nevertheless, as the engines were started and our huge jet took off, I realized it was too late to turn back.

Conversation was next to impossible as we flew westward over the Pacific Ocean. At noon the sergeant passed out box lunches. After that I was invited to look at the flight deck, elevated a few steps above the floor of the plane. From there I got my first view of Eniwetok Atoll, the large roughly circular edge of an old volcano with a string of small islands dotted along its western side. The large lagoon inside the circle was a beautiful light blue, paler in the shallow places; outside the circle to the east the water was a deep dark blue where the ocean floor dropped almost vertically to the Pacific deep.

I was greeted upon landing by Dr. Don Nelson, Ralph Nelson, and Rod Allin. We soon met the commanding officer, and I was shown to my room, small but comfortable and with a bath shared by an officer in the adjoining quarters.

I met the graduate students, Bob Johnson and Greg Pettiger, in their large workshop-laboratory and there saw the two submersibles that were to be used in Dr. Nelson's underwater research with gray reef sharks. This is the species involved in the majority of attacks on people in Australian waters. The bodies of the submersibles were similar in shape to a shark; the one to be used by Dr. Nelson was painted blue; and the one I was to pilot, yellow. The strong fiberglass hulls would withstand a shark attack, and so would the Plexiglas observation windows. Each craft had two small propellers on either side near the Plexiglas bubble; these were powered by batteries encased in waterproof tubes and were operated by pulling a lever up or down. Rudder and elevators in the rear were controlled the same way as in an airplane. A tank to supply air to the driver was positioned in a rack and could be detached in an emergency to allow the operator to abandon the sub and swim to the surface.

At dinner in the mess hall that evening we met many of the people involved in the decontamination work, including a team of doctors from Bethesda, Maryland, who were on hand to make one of their frequent inspections for contamination of both the area and the personnel involved. Their inspections were thorough and they looked forward to the time when the inhabitants could return to their islands and live as safely as before the atomic bombs were exploded.

Next day, after I had checked out all my scuba equipment, we decided to postpone our outing in the submersibles and instead to take the power boat across the lagoon and dive from the outside wall, where Dr. Nelson had previously seen a goodly number of gray reef sharks. When we arrived we dropped anchor in thirty feet of water on the narrow shelf of the top of the reef, scanned below, sighted sharks, and slipped quietly into the water. Two men remained at the surface and the other six descended the reef wall to about sixty feet and started filming. Dr. Nelson and I stationed ourselves with our backs toward the protection of the wall; we each carried an aluminum pole with a powerhead at the end away from us. The other divers were similarly equipped. Ralph and Rod were nearby with their cameras; as the sharks swam near us, they would film the activity.

We counted fourteen sharks in all. At first they kept away from us, but as the minutes passed they became more curious and swam in closer. Two sharks moved slowly toward us from our right, following along the wall. Don, on my left, and I watched them approach. When about thirty feet away, one turned away from the wall. The other continued on, however, and when he got as close as about ten feet, I moved my pole until the powerhead was pointed directly at him.

Seeing it, this shark also turned quickly from the wall and swam away from us.

We continued to film for about forty minutes and as the sharks continued to approach closer to us, we decided we had accumulated enough footage and began our ascent. While this was being photographed by Ralph and Rod, a long thin row of slowly swimming barracudas passed just over our heads. We paused to let them go by, and I counted thirty-two before they had moved out of view.

We continued our ascent. On the way I reached the anchor rope and pulled myself along to the surface, where I handed up my weight belt and tank and swung myself into the boat. When we were all aboard, the anchor could not be dislodged from the coral. Bob Johnson volunteered to go below and find out why. Of course he took along a powerhead; two men were watching nearby and the rest of us from the surface. Keeping a sharp eye on the seven or eight sharks near him, Bob managed with his free hand to dislodge the anchor and then surfaced without further incident. Back in the boat, he told us that other sharks were beginning to concentrate on his position right up to the time he managed to free the anchor.

For several days we continued to film marine life in the lagoon. At one place in about thirty-five feet of water, we sighted a Japanese fighter plane resting eerily on the bottom, right side up. We swam all around it and found the bullet holes that must have caused it to crash. In the cockpit were parts of a human skeleton—a grisly reminder of the horrors of war.

One evening the commanding officer asked me to give a short talk to the staff of another camp nearer to the contaminated area. I agreed and was flown there in a helicopter. I took along my still camera and was able to get some beautiful shots of the narrow ridge of the atoll with the waves breaking in the shallow water. At one point we flew over a deep, dark-blue, round hole caused by the blast of an atomic bomb. I was told that this was where the radioactive material was being buried; when the job was completed, a thick covering of concrete would hold the hot matter intact there under the surface. On another evening Ralph and I spoke to the service personnel in a small, open-sided chapel about our work for "Mutual of Omaha's Wild Kingdom" and about the filming we were doing with Dr. Nelson.

At last the day came for filming shark aggression against Dr.

Nelson in his blue submersible and me in the yellow one. I had practiced with my sub and could control it fairly well. When on location at a depth of about thirty-five feet, Dr. Nelson inserted a cigarlike cylinder into the mouth of a twelve-inch fish; he waited until a shark swam by his boat, then dropped the fish out, letting it settle through the water, where it was snapped up by the shark. The cylinder was a telemetry device designed to send out a radio signal under water. This signal could be picked up by a receiver carried in Dr. Nelson's submersible. Nelson hoped the signaling animal would lead him to an area where there were more gray reef sharks for his studies about aggression and attack.

The blue sub followed the signal and the yellow sub followed the blue. Our support boat moved along on the surface. Finally Dr. Nelson caught up with his shark and must have been happy to see a group of hungry-looking gray reef sharks cruising and circling about. Amid them, I saw the blue sub come to rest on the sea floor. The support boat observed this too, and Ralph and Rod in their scuba gear and carrying cameras slipped into the water and swam down to the sub. I, too, circled about and watched as our support boat came to a stop on the surface and dropped anchor nearby.

Signaling to us, Nelson started his propellers and took off in the wake of a six-foot shark. The two photographers followed him and so did the yellow sub. Nelson continued to pursue the shark, staying close behind it and turning each time the shark did. This went on for a short time; then suddenly the shark went into the agonistic attitude, arching its back, dropping its pectoral fins low, and swimming in an erratic manner, rolling from side to side. Then, in a flash, it turned and made a swift attack on the blue sub, striking it along one side with an open-mouthed bite. The teeth tore off the blue paint in streaks, baring the white fiberglass beneath in two rows for the upper jaw and two for the lower. Ralph and Rod recorded the whole onslaught on film.

Nelson, undeterred, brought his sub about to continue following the shark. The two photographers were close behind. After only a short chase the shark again assumed the agonistic attitude and then, with great speed, made a second attack on the sub. This time it went for one of the revolving propellers and bit off two of its blades. This incapacitated the sub and as Nelson cut the power it slowly sank to the bottom. Again Ralph and Rod recorded the action on film.

Nelson calmly detached his air tank from its rack, shouldered it, and swam out of the battered submarine. With the help of other scuba divers the sub was manhandled into shallower water, whence it could eventually be returned to the laboratory for repairs. I rejoined the other divers, much impressed by the risks and pains today's scientists are willing to take to probe and better understand the behavior of sharks.

A striking example of modern science's intrepid approach came when we joined Dick Johnson, a graduate student of Nelson, who was studying shark behavior at Rangiroa Atoll, about 300 miles northeast of Tahiti in the South Pacific. Ralph Nelson and Rod Allin had gone ahead of Carol and me to make the preparations for filming another underwater episode. Kia Ora Village, where we stayed on this beautiful atoll, was the image of what one would expect a native-style, South Seas resort hotel to be. There was an open-sided, thatch-roofed dining area facing the blue lagoon. An ancient dugout canoe decorated one side. A boardwalk led over the water to a round open-air pavilion where drinks were served. Thatched-roof cottages were scattered among the coconut palms, each with a porch and a large picture window facing the lagoon. The outside of our cottage was rustic, but the inside was modern—electric light, good beds, inside plumbing and a hot and cold shower, rugs, curtains, and draperies. The food was excellent. A conch shell horn announced meal time. Compared with the dusty, dirty, cold, primitive places I had stayed when filming "Wild Kingdom," the word to describe Kia Ora was "paradise."

The clarity of the water at Rangiroa was phenomenal, and the water was warm. There were glades of clear white sand, coral heads, and reefs with a great variety of fish. Dick Johnson's yacht was anchored at the harbor. He had been at Rangiroa for some time studying sharks. White-tipped reef sharks were numerous in the lagoon. Gray reef sharks were present too, and Dick kept his eye on them when they showed up, and slowly backed to a coral head to reduce the chance of an attack from behind. He told us of a spot nearby where he had been conditioning white-tipped reef sharks to take food from his hand.

To film this activity, Ralph, Rod, and I dove with Dick from a rubber boat anchored by the shark-feeding area. Dick took with him a supply of dead fish, each about fourteen inches long. He indicated a spot where he would lie on his stomach on the bottom with his arm extended and holding a fish in his hand. I was to be about level with his feet, and a little to one side, also lying on my stomach. Ralph and Rod took their positions for the filming, each resting on the white-sand bottom. Dick took a fish out of a plastic bag and held it forward and higher than his prone body. We did not have long to wait before a white-tipped reef shark swam toward Dick, but about eight to ten feet above the bottom. It circled and then swooped in lower, mouth gaping, and took the fish. Dick didn't even flinch. The shark swam off swallowing the fish.

Dick looked to Ralph for instructions and received a circled forefinger and thumb to indicate the photographers had recorded the whole episode.

Ralph changed his camera position a little and nodded that he was ready for a second run. The shark did not circle so wide this time, and again both photographers were happy. The third and fourth time more than one shark came in. Dick continued to feed them one at a time.

I was amazed that wild, free-swimming sharks could be habituated to hand feeding. It was certainly contrary to most of the stories you hear about sharks.

Dick Johnson, whose book, *The Sharks of Polynesia*, appeared in 1978, is still in French Polynesia, and he has not yet been bitten by a shark.

In 1981 I was happy to have an opportunity to dive with another graduate student of Don Nelson, Peter Klimley, who was diving with Nelson in the Sea of Cortez to observe scalloped hammerhead sharks. To film this activity, the "Wild Kingdom" crew traveled to La Paz in Baja California. Off that coast Don and Klimley had found a submerged seamount where at depths of from fifty to eighty feet the hammerheads congregated in a school of 150 or more.

As I walked off the jet plane at La Paz Airport, Peter Klimley was waiting for me with Ace Moore, associate producer of Don Meier Productions. We drove the short distance into the city, a city I could not recognize. When I had last been there fifteen years earlier it was a very small place, but I now saw a thriving city with skyscrapers. The hotel was air-conditioned, spacious, and well managed. The people behind the check-in desk expected me to speak impeccable Spanish, as the soundtrack for the "Wild Kingdom" they see has a Spanish voice dubbed in. Fortunately, they all spoke good English.

Around the dinner table that evening I met all the research staff involved in the hammerhead project and learned of their jobs and interests. Some of their work had already been filmed. Peter Klimley and Don Nelson had been researching hammerhead behavior for two years, and an hour of videotape had been accumulated. Still pictures had also been taken using twin underwater cameras on a bracket, and from these pictures they were able to determine that the length of the sharks was in the eight to twelve foot range.

Manta rays with "wingspans" of twelve to fifteen feet were abundant, and whale sharks, largest of the elasmobranchs, creatures reaching whalelike porportions of fifty feet in length, cruise these waters feeding on plankton. Because of its size and feeding habits, the whale shark is easily approached by divers, and some have hooked a ride by holding on to a dorsal fin.

We sailed the following morning from a dock within walking distance of our hotel, aboard a diesel-engine boat built in Cuba and traded to Mexico. After leaving the harbor our speed was increased, for we had a seven-hour cruise before reaching the seamount. The captain set a course to the northeast, and I settled in with my equipment, but as I was a newcomer I found all the good places occupied. The only spot left for my gear was the floor space in the head.

When we approached our destination Peter pointed out two notches in a long row of mountains on the western horizon. By heading on a course precisely between these notches, the seamount could be located. A couple of the divers, by snorkeling on the surface, found the exact location. We anchored there.

Rod Allin was one of our underwater cameramen; Howard Hall, from San Diego, was the other one. Howard had come to us highly recommended as a free-lance photographer. I was to learn that he was an ardent conservationist, a board member of the Living Ocean Society, and had filmed the killing of dolphins on Iki Island in Japan. That had been a dangerous experience, as he had been threatened with a long-handled lance used to kill the dolphins.

He was just the man to join Peter and Don as they free dove to sixty or eighty feet to tag the hammerhead sharks and record their behavior with a video camera. These scalloped sharks are so-called because of the scalloped formation of the fronts of their grotesque, flat heads, on each end of which the eyes are placed. Peter and Don had at first tried to approach the large school of hammerheads in scuba gear, but each time they came close the sharks moved away, disturbed by the air bubbles rising from the equipment. After many attempts to get close to the sharks, they realized they could not do so in scuba gear. Being young, athletic, experienced swimmers and dedicated scientists, they decided to leave the air tanks on board ship and free dive down to the sharks. This worked so well that they were able to approach the sharks and finally to come close enough to attach identification markers and cigar-sized telemetry devices (radio beepers). The markers and beepers were affixed from the end of a ten-foot pole into the tough skin of the sharks' backs. They enabled the researchers to identify a shark as one from the seamount and to follow it to determine where it went when it left the seamount, as all the sharks did at night.

The video camera in its underwater housing was the size of a small suitcase and had to be pushed through the water just in front of the diver. Each scientist could hold his breath about a minute and a half. During that brief period, one at a time, they would leave the surface, wearing only mask, snorkel, and fins, swim down sixty to eighty feet, tag a shark, turn away, and swim back to the surface.

Howard Hall free dove with them and filmed this procedure. In order to do so Howard preceded the diver into the water, photographed his descent to the sharks, the swift tagging, and the return to the surface. Howard could hold his breath an astonishing two minutes.

To observe and film the tagging operation from my point of view, Rod and I dove in scuba gear to the top of the seamount, fifty feet below the surface and ten to thirty feet above the sharks. Remaining still, with my knees resting on coral at the summit of the seamount, I could look around me and see the myriad fish in the clear blue water. A manta ray swam past in its characteristic easy, unhurried "flight" action. Just beyond him was a long line of small tuna in a column of movement that streamed past for several minutes. Looking down to my right, I could see the great school of hammerheads, all moving casually in the same direction, each a distinct gray form against the indigo background of the big deep. All were in the range of ten to twelve feet in length and swam continuously as some sharks must in order to breathe. They had no destination, just swam counterclockwise around the seamount.

My thoughts ran back to my boyhood when I had made a crude underwater breathing apparatus from a length of garden hose, one end threaded through a hole in a floating board, the other end in my mouth. A rock in each hand kept me underwater as I walked across the bottom of an old strip mine. How different to be here in the Sea of Cortez sixty-six years later, fifty feet below the surface, resting in comfort, breathing normally, and watching a school of about 150 scalloped hammerhead sharks swim to within forty feet of me. I felt no uneasiness, no fear—only gratitude for this privilege—and I sent a mental message of thanks to Jacques Cousteau for his part as coinventor of the aqualung that made this experience possible.

Howard Hall, also in scuba gear, was hanging in the water slightly behind and above us. Looking up through the clear blue water, I watched the two figures on the surface. I knew one was hyperventilating, breathing deeply eight or ten times to get as much oxygen in the blood as possible. Then I watched as he surface dove and with slow, rhythmic grace flippered his way down toward us. He seemed not to hurry, but his larger-than-normal flippers were moving him at a good speed. Pole in hand, he passed our level about thirty feet in front of us and continued on down to the hammerheads. A quick thrust with the pole, a slow, graceful body turn to the vertical position, and with the same measured flipper movement he rose from the dark blue background of the deep to pass us on his unhurried ascent to the aquamarine blue of the surface. It was beautiful and graceful and wonderful to watch.

On the third day we awoke to a dark, threatening sky. The wind was blowing hard, and an ominous dark cloud loomed in the south. A larger ship had anchored near us the previous day, and just before its crew hoisted anchor to depart, they advised us that their ship-to-shore radio had warned of a hurricane. After a quick conference, we decided to return to La Paz. In pitching waves and rain we cruised past islands, staying on the leeward side as much as possible until late afternoon when we entered the calmer waters of the Bay of La Paz. We waited two days for the storm to wear itself out, which finally it did by crossing the Gulf of California and petering out over the mainland of Mexico.

When we returned to the seamount we found the sharks still in residence, so again we entered the water to continue filming. Once more I sat on the submerged mount, waiting for Peter to get ready for a dive with the video camera. A manta ray with a "wingspan" of twelve to fifteen feet passed just in front of me, traveling with an action that seemed more like flying than swimming. I swam forward for a better look, but could not overtake the manta. In the following days we saw many other manta rays. On one occasion a big one was headed toward Don Nelson. Don swam slowly onward, thinking the ray would see him and change course. Instead, it came close enough for Don to reach out and touch its head. Startled, the manta ray turned up instead of down and collided with Don, hitting him in the face. Fortunately, it was a glancing blow and did no real injury.

On other occasions one or more of the divers, swimming just above a slowly moving manta ray, hooked a ride by taking a handhold at the front of its head.

Several whale sharks were also seen. A diver hooked a ride on a youngster about twenty feet long by holding onto its dorsal fin.

There is still much to be learned about sharks through the methods of modern technology. Investigative scientists such as Peter Klimley and Don Nelson will further our understanding of these interesting and primitive creatures of the sea.

28

Giant Lizards of Komodo and a Dream Fulfilled

While on my learning trip to the New York Zoological Park back in 1927, when I was a young man working at the Saint Louis Zoo, I spent some time with Dr. Raymond Ditmars and also visited for several days at the American Museum of Natural History. There Dr. G. K. Noble, curator of herpetology, was most courteous. He gave me time to talk reptiles and helped me to understand museum research and the collecting of specimens. He showed me the museum's catalogue filing system and the rows of preserved specimens available to scientists for a variety of studies. On one occasion he gave me the first copy I'd ever seen of *Natural History*, the magazine for museum members. That issue contained a stimulating article by Douglas Burden describing his trip to Komodo Island, near the Flores Islands—a string of Indonesian islands that stretches eastward from Java to Flores Island north of Australia.

Burden had chartered a boat in Bali to take him to Komodo to study the world's largest lizard—the Komodo dragon lizard. He described nine-foot-long monsters and wrote of his difficulties in capturing two of the big lizards for the New York Zoological Park. Included in the article were photographs of the lizards and their imprint. After reading Burden's account of his experiences and difficulties on Komodo, I knew that a similar trip would be one of my goals for my own future expeditions.

Some years later Pete Miles, who was an animal keeper at the Saint Louis Zoo, resigned that job and became involved in the animal-dealing business. One of his trips had taken him to Surabaja, Java, and naturally he went to the zoo, where he saw several large specimens of the Komodo monitor lizard. Pete was able to buy the largest of the lizards, and with it safely in a big crate he set sail for San Francisco. From there he came to Saint Louis, and George Vierheller allowed him to house his animals at the zoo, where they could be cared for while Pete contacted other zoos about their sale.

We had only one cage suitable for the gigantic lizard—a circular barred cage near the lion house. Pete agreed to let the lizard go on display there.

While moving the animal in, I weighed and measured it. Its total length was ten feet two inches; it weighed 365 pounds. The huge creature made a fine exhibit and Pete got good publicity about his collection of animals, all the press stories featuring the Komodo monitor lizard.

The Chicago World's Fair was on in that year (1932), so Pete went to Chicago to see about selling his lizard to Frank Buck, who had an exhibit of wild animals there. He showed pictures of the lizard to Buck and his business manager, and was offered $20,000 for the animal. Pete told Buck he must return to Saint Louis to take care of his lizard and other animals and that he would telephone the next day with a firm yes or no.

On the train back to Missouri Pete thought about the large offer and after considering various angles of the transaction decided on a no, on grounds that if it was worth $20,000 to Frank Buck to exhibit the lizard at the fair it was worth more than that to Pete to exhibit it around the country.

The morning Pete arrived at the zoo from Chicago, all set to make a fortune exhibiting the largest lizard in the world, I had to tell him that the monitor had died that morning. Pete could hardly believe the news. The autopsy showed pneumonia. What a blow for Pete, and for all the people who would have seen the giant lizard if it had lived!

A few years later, in a book I wrote titled *Animal Faces*, I showed a portrait of Pete's Komodo dragon lizard and stated its length and weight. It remains the largest specimen ever recorded.

When "Wild Kingdom" was beginning I listed Komodo as a possible story idea. In 1970 we learned of Dr. Walter Auffenberg's recent studies of the Komodo monitor lizard. We found that he had spent a year on Komodo Island researching the big lizard's habitats and physiology, and we contacted him at the University of Florida in Gainesville. A plan was worked out. Dr. Auffenberg agreed to go with us to Java, taking along a graduate student who would do a research project on the blood chemistry of the lizards. He would also investigate their saliva to try to determine the causative organism of the infection that always follows a bite by one of these lizards.

Walter had learned the Indonesian language while there and had employed as his assistant an Indonesian student of zoology, Putra Sastrawan. He would be contracted to go with us as well.

When the appropriate time arrived, Warren Garst, our photographer, and I flew from East Africa to Jakarta via Bombay, India. In Jakarta we met Dr. Auffenberg and Tom Allen and Rod Allin, who had flown in from Chicago and Florida. We stayed a few days in Jakarta securing various permits, then made a short plane flight to Bali, where we spent several days arranging for the charter of an Indonesian coastal vessel that often went to Komodo on official business. We also got together the equipment and supplies we would need to be self-sustaining on Komodo for four weeks. The equipment included mats for our sleeping shelter, blankets, wire, tools, nails, cooking utensils and tableware for our meals, buckets, lights, a variety of fresh vegetables, fruit, nuts, dried foods, rice, and canned goods.

It was a privilege to have Walter Auffenberg on board with us as we steamed out of the harbor at Bali and headed eastward. Walter had studied Indonesia well and kept us informed of our whereabouts. He told us names of the islands we were passing and something of their zoogeography and inhabitants. At one island he mentioned that when the Dutch were first establishing their colonies in the East Indies their ships were sometimes pirated by the fierce natives of the Bogey tribe who boarded the Dutch ships in the night, killed the crew, and plundered the vessels and their cargoes. It was these pirates who gave us our "bogeyman."

Our accommodations on the steamer gave us only deck space, as the cabins below deck were for the crew. Toward late afternoon an open fire was started in a crude stove on the afterdeck and a simple dinner was cooked. We opened a couple cans of food as well and finished off with some fresh fruit.

As darkness fell, we could see the fires on the islands we were passing. We rolled up in our blankets on the foredeck and slept fitfully during the night. Toward the afternoon of the second day we sighted the north coast of Komodo and rounded the east side, where we photographed the tall spires of the throat of an inactive volcano still standing after the outside portions had been eroded away. This had been well described by Douglas Burden.

Continuing on around the island, we turned into a wide harbor and saw the native village on the shore and people running back and forth pointing in our direction. Two boats put out from shore, and soon they were alongside and Walter was calling out the names of his friends and greeting them. He received a warm welcome. When the villagers came on board Walter spoke to them of our plans and told them that we would be staying at the same campsite he had used before.

Our gear was transferred to the smaller boats, and we set out for what was to become our home. Everyone helped carry the equipment to the site, and we set about immediately erecting a framework for our sleeping quarters and putting up mats as walls between the bed areas. However, we left the side facing the sea open to take advantage of the breezes.

Walter's old table was brought in by people from the village. Not wanting to miss the excitement of our arrival, they had walked a mile carrying it and had also brought a gift of some coconuts. Walter arranged for a camp boy to gather wood and to boil the water that would be brought daily by boat, in clean jerry cans, to our beach. A large rough-barked tree shading our campsite became a rack on which we hung some of our sacks of food. One net bag was filled with brown-sugar cakes for a quick energy pickup. One of the natives spied them and helped himself. Others followed, and since we were busy setting up camp we did not notice the rapid depletion of sugar cakes until they were all gone. Walter explained to us that the natives had not stolen the sugar; in their society it was customary and expected for a guest to help himself from his host's larder.

After that we were more careful about putting things away, but local custom continued to prevail. Walter told us it was one reason he had wanted the camp away from the village.

Before dark we walked to the beach and had a refreshing swim in the cool water. It felt good to wash and change into clean shorts before dinner.

A government official had come on the ship with us, and the next morning the craft continued on to deliver him to one of the Flores Islands. We did not expect to see our ship for another two weeks. But two days later it was back and anchored in the bay; those aboard started fishing. Walter went out to talk with the crew to make sure they understood the date we planned to start back to Bali. They informed him they had decided to stay until we were ready to go. That was welcome news, for we had wondered what we would do if one of us became sick or broke a leg or was bitten by a large lizard.

We set to work the day after our arrival to look over the area Walter knew so well. We followed a path that led to a dry riverbed, and Walter searched for a nine footer he called "number nine," who had lived on the hillside to our right. We all spread out and climbed the hill, but we saw no monitors. We did, however, reach a place of great beauty that overlooked the whole valley and bay and gave Warren the master scene he needed to display the quality of the island. A long walk on the beach showed him a likely place to set his blind and photograph wild boars and native deer, the prey animals of the big lizards. A spot in the dry riverbed was picked to set the bait and trap for *Varanus komodoensis*. Arrangements were then made with the chief of the village to sell us a goat; this was killed and hung at that spot. A blind was set up nearby so that anything approaching the bait could be photographed without danger of spooking it.

In a couple of days lizards began to appear around the bait, small ones three feet long at first. These were photographed and then caught with a slip noose, a process which was also photographed. All captured animals were measured, scale clipped, and weighed, and blood and saliva samples were taken. They were then released, and some were soon at the bait again and were caught a second time. We could tell this because of the scale clipping, which gave a coded number to each individual lizard without in any way harming the animal.

One day we photographed our largest Komodo lizard, caught it, weighed it, and measured it as longer than seven feet. When this monster came to the bait, a three footer took to a tree and climbed about thirty feet up the trunk. On another occasion a six footer was seen swimming toward Komodo Island from a small island about a mile offshore. Tom and Rod grabbed their masks and snorkels, jumped in a dugout canoe, and Rod photographed Tom underwater, following the lizard, and recorded the animal's emergence from the sea when it left the water and walked ashore.

From camp it was about a mile down the beach to the blind for deer and wild-boar photography. We went there frequently, and Warren or Rod spent long hours photographing from the blind always hoping for an exciting event, such as a large lizard catching, killing, and eating one of the prey animals. I too went along when I was not needed on camera and sat quietly in the blind to record with my still cameras the scenes we were getting on the Arriflex 16-mm movie cameras.

Dr. Auffenberg told us that he had tried to verify the many reports that Komodo monitor lizards had attacked large animals, including man. He had traveled to some of the nearby islands where these large lizards also lived, and talked to eyewitnesses of such encounters. One such was the killing of a water buffalo on a farm. The lizard approached the buffalo, which was grazing in a field, and made a direct attack from the rear, grabbing a hindleg with its serrated, sharp teeth, cutting the large tendon of the leg and incapacitating it. Another quick attack put the other hindleg out of action. The lizard then proceeded to help himself to the flesh of the crippled buffalo. Horses were likewise killed and eaten by the large lizards.

Two instances were recorded of attacks on the people of the village on Komodo. A man and his two sons, one twelve and the other fourteen, went up the mountain to get firewood. They were cutting and piling the wood in a clearing when, looking up, one of them saw a lizard about seven feet long with his head held high watching them from the edge of the clearing. They were not alarmed, as they often saw these lizards around the island. The lizard then began walking toward them, and soon they realized that it was in fact increasing its speed and was after one of them. They all three started to run, but the fourteen-year-old stumbled on a root and fell to the ground. Quick as a flash, the lizard was upon him and, with a quick bite, took off the right side of his buttocks. The father and other son charged the lizard with sticks, hitting it to drive it off, but before they could do so the lizard had bitten off the fleshy part of the boy's leg. The lizard then retreated, and father and son picked up the injured boy and started back to the village. He was bleeding badly, and they had no way to staunch the blood. Just as they brought the boy into the village, he collapsed and died.

Another attack involved two men cutting wood on the mountain. One became ill, so the other left him lying on the ground and went to the village for help. When he returned to the location on the mountain, all that remained of the sick man was his head.

During the third week we ran low on food in our camp. We bought fresh fish from the villagers, who had a large fish trap in the bay. But without cooking oil the grilled fish were dry and unappetizing. We ran out of breakfast cereal and powdered milk. We exhausted our stock of canned and dried foods. For the last week our larder contained only cabbage heads and rice. Our meals took on a sameness—boiled cabbage and boiled rice. We had a little salt left for seasoning. The village could provide no more goats, and the supply of coconuts had run out. Our food wasn't gourmet dining, but it was sustaining. Rod Allin had for some time been bothered with stomach ulcers. We were amazed when, upon arriving back in Bali, we heard Ron say, "By golly, my ulcers are gone."

One day, walking along a trail left by deer and wild boars, we noticed a small grayish cobra climbing a small bush. A camera was quickly set up and I walked in near the snake and put on my glasses. This type of cobra, like the red-necked cobra of Africa, has a rare ability: It can spray its venom. And once in the eyes of an adversary the venom can cause blindness or even death through absorption. I was taking no chances. As I moved near the snake, it spread its head and turned its nose in my direction. I slowly took my binoculars off my shoulders and holding them up by the strap moved them closer to the cobra. As I expected, he changed his position to face the nearest moving object, and as the binoculars continued to move he opened his mouth and ejected his venom in a fine spray. At the same time there was an audible hiss as air expelled from the slitlike opening of his trachea in his lower jaw carried the venom even further than he could squeeze it from the openings in his fangs. I saw the tiny drops hit the lenses of my binoculars. He was indeed aiming at the "eyes." Again and again I swung my binoculars so they would pass within a foot of the cobra, and each time he spat his venom at them. A closeup taken by our movie camera showed the tiny crystalline drops on the glass of my binoculars. I did not take my own glasses off until I was safely out of range of the cobra.

During those days on Komodo I often walked alone along the shore. And each time I would stop and stand looking at the beach, and often at the foot and tail tracks left by a big lizard who had also passed that way. Then my mind would run back in time to the American Museum of Natural History in New York, where I had first learned of Komodo and its large lizards from Douglas Burden's article in *Natural History*. With a feeling of accomplishment I reflected that now I, too, had come to Komodo to record the activities of the largest lizards in the world on movie film that would be seen by millions of people not only in the United States, but in some forty other countries around the world. It was a contribution I could be proud of.

After one of these feelings of elation, I heard my own voice saying, "Marlin, you are walking along a beach on the island of Komodo, and there are the tracks of a Komodo dragon lizard to prove it; and there, washed up from the sea, is the shell of a chambered nautilus, a distant relative of the squid; and there is the fragile shell of another relative, the paper nautilus. You have recorded these wonders all over the world. Marlin, you are a very fortunate fellow."

Index

CHRISTIAN HERALD ASSOCIATION AND ITS MINISTRIES

CHRISTIAN HERALD ASSOCIATION, founded in 1878, publishes The Christian Herald Magazine, one of the leading interdenominational religious monthlies in America. Through its wide circulation, it brings inspiring articles and the latest news of religious developments to many families. From the magazine's pages came the initiative for CHRISTIAN HERALD CHILDREN'S HOME and THE BOWERY MISSION, two individually supported not-for-profit corporations.

CHRISTIAN HERALD CHILDREN'S HOME, established in 1894, is the name for a unique and dynamic ministry to disadvantaged children, offering hope and opportunities which would not otherwise be available for reasons of poverty and neglect. The goal is to develop each child's potential and to demonstrate Christian compassion and understanding to children in need.

Mont Lawn is a permanent camp located in Bushkill, Pennsylvania. It is the focal point of a ministry which provides a healthful "vacation with a purpose" to children who without it would be confined to the streets of the city. Up to 1000 children between the ages of 7 and 11 come to Mont Lawn each year.

Christian Herald Children's Home maintains year-round contact with children by means of an *In-City Youth Ministry*. Central to its philosophy is the belief that only through sustained relationships and demonstrated concern can individual lives be truly enriched. Special emphasis is on individual guidance, spiritual and family counseling and tutoring. This follow-up ministry to inner-city children culminates for many in financial assistance toward higher education and career counseling.

THE BOWERY MISSION, located at 227 Bowery, New York City, has since 1879 been reaching out to the lost men on the Bowery, offering them what could be their last chance to rebuild their lives. Every man is fed, clothed and ministered to. Countless numbers have entered the 90-day residential rehabilitation program at the Bowery Mission. A concentrated ministry of counseling, medical care, nutrition therapy, Bible study and Gospel services awakens a man to spiritual renewal within himself.

These ministries are supported solely by the voluntary contributions of individuals and by legacies and bequests. Contributions are tax deductible. Checks should be made out either to CHRISTIAN HERALD CHILDREN'S HOME or to THE BOWERY MISSION.

Administrative Office: 40 Overlook Drive, Chappaqua, New York 10514
Telephone: (914) 769-9000